S0-AKJ-868

"What?" said Jack.

"Hercules is a myth to people in *this* time," said Annie. "But in Roman times, lots of people believed he was real. So, since we were in Roman times, he was real to us."

"I don't know..." said Jack.

"Did you ever hear the saying?" said Annie. "*When in Rome, do as the Romans do.*"

Jack laughed. "Yeah." He looked up at the sky. "Thanks, Hercules," he said softly, "whatever you are."

"Jack! Annie!" their dad called from their front porch. "Time to go!"

"Oh, brother, I forgot," said Jack.

"Yeah, I hope nothing exciting happens on *our* vacation," said Annie.

"Yeah," said Jack. "I hope it's really, *really* boring."

"Hurry!" their dad said.

"Coming!" they called.

Then they took off running for home—and a restful vacation.

Plant Growth

Plant Growth

Michael Black, Ph.D.

Lecturer in Plant Physiology, Queen Elizabeth College, London

and

Jack Edelman, Ph.D., A.R.C.S., F.I.Biol.

Professor of Botany, Queen Elizabeth College, London

Harvard University Press
Cambridge, Massachusetts
1970

First published 1970

SBN 674 67262 3

Printed in Great Britain

Preface

One of the foremost questions facing the biologist is how organisms grow and develop into what they are. This problem, which is at the frontiers of knowledge, now challenges the modern botanist and he, together with the molecular biologist, biochemist and biophysicist, is increasingly turning his attention to its solution.

But despite rapidly growing research, this subject which is of such fundamental and economic importance receives only cursory treatment in biological teaching. This book sets out to present a comprehensive account of the growth and development of plants, in an attempt to remedy the deficiency.

We have emphasised processes, and have attempted to explain how they happen, trying to avoid a catalogue of descriptive factors. After all, the plant does not separately categorise its growth processes, for instance enlargement and differentiation, nor its internal controls such as the individual hormones, nor, again, the external factors of light, temperature, etc. Instead, plant growth is an integrated process which occurs by the interaction of all these.

We hope that this book will be a useful introduction to the subject, especially to sixth-formers, to students at technical, teachers' training and other colleges of further education, and to undergraduates.

M. B.
J. E.

Acknowledgements

Plate 1 Photograph by courtesy of Dr E. H. Newcomb.

Plate 2 (*a*) Photograph by D. & H. Adamson, provided by courtesy of Professor R. D. Preston.
(*b*) Photograph by R. K. White, provided by courtesy of Professor R. D. Preston.

Plate 3 Photographs by courtesy of Dr E. G. Jordan.

Plate 4 Electron micrographs provided by courtesy of Dr Y. Lemoine.

Plate 5 Photograph by courtesy of Plant Protection Ltd; kindly supplied by Professor L. Audus.

Plate 6 (*b*) Photograph and information by courtesy of Sussex Nurseries Ltd., Rustington, Sussex. Phosfon is registered trade mark of Mobil Oil Corporation, New York, U.S.A.

Plate 7 (*a*) Photograph by kind permission of Dr S. H. Wittwer, Michigan State University.
(*b*) Photograph by courtesy of Professor P. W. Brian and I.C.I. Ltd.

Plate 8 Photograph kindly supplied by Professor Anton Lang.

Plate 12 Photograph kindly supplied by Professor F. C. Steward; published originally in *Science*, **143**, 20, 1964 and subsequently reproduced as fig. 10–6 in *Growth and Organisation in Plants* by F. C. Steward.

Plate 14 Photographs kindly supplied by Professor F. Salisbury; previously reproduced in *The Flowering Process* by F. Salisbury.

Plate 15 Photographs kindly supplied by Dr Bruce Cumming.

Plate 16 Autoradiographs provided by courtesy of Professor K. Mothes.

Contents

List of Plates

Introduction

The most obvious thing that plants do is to grow. This is such a natural background to our existence that the astonishing changes which occur may pass with little comment, but when a seed grows into a mature plant, it alters remarkably in size and shape. Roots, stems and leaves appear and during maturity, organs concerned with reproduction, the flowers, fruits and seeds, are formed and grow. These obvious morphological changes which accompany growth in size are called *development.*

As plants are made of cells, we can infer there are changes in the number and types of these, giving rise to the developing tissues. This changing cellular pattern is called *differentiation.*

Growth does not occur haphazardly but is *co-ordinated*: for instance, the shape of mature leaves is usually constant for a given species; flowers are produced on shoots, not on roots. This co-ordinated nature of growth implies the functioning of extra-ordinarily precise *controls*: in an acre of wheat hardly a single plant can be found to have 'gone wrong' – no mass-production line could claim such success! These controls are sensitive to the environment. For instance, in temperate and arctic climates the growth of an annual is very obviously affected by the seasonal cycle; if a plant is grown in the dark it looks quite different from one grown in the light; shoots grow away from, and roots towards, the gravitational pull of the earth, and there are innumerable other examples. Plants are much more affected by their environment than are many animals which develop a more standardised internal environment of their own, and some of this book will be taken up with the discussion of the effect of external factors on growth and development.

We must also consider in great detail the internal factors to which growth is subject as they are also of major importance. Thus plants of different species starting growth in the same habitat grow differently, for example, the bluebell and the oak, and this is due in part to differences in internal control, which are genetically determined. One organ can greatly affect the development of another, a phenomenon called *correlation*: for instance, the growing apex of a stem often suppresses the development of the lateral buds, an effect called apical dominance.

Paradoxically we must also consider as part of our subject total suppression of growth, sometimes in apparently favourable conditions. This phenomenon of *dormancy* is shown by some seeds, and also by winter buds of trees. *Senescence*, or decline of growth leading to death, is also obviously a part, albeit the termination, of the overall developmental process.

Growth appears always to be an irreversible process, and development from simple to complex organisation at all levels, followed by decline and death, is a characteristic feature of all living organisms.

Why and how do all these manifestations of growth and development occur? How are they controlled? How is the cell, tissue, or organ aware of, and influenced by, various external and internal factors? These questions strike at the core of biology – the nature of life itself. It is generally accepted that no one of the organic substances which constitute living organisms can explain the phenomenon which we call life, and we must seek our explanations of living processes not in any particular substance or 'essence' but in the complexity of the organisation of structures and chemicals. A good analogy is a motor-car in which the complex organisation, co-ordination and control of simple components enable complicated events to take place. In recent years a great revolution in biological research has occurred, away from merely classifying and observing living organisms to finding out more about the complex interacting events which enable them to grow and function as they do. The new subject of molecular biology, for instance, is one avenue by which we are trying to explain the control of growth processes.

To summarise, the main processes we will investigate in this book are development – the obvious morphological changes which occur during growth, – brought about by differentiation of simpler cells and tissues into more complicated ones. The development of individual organs affects that of others, they are correlated, and growth as a whole is co-ordinated to give the mature organism which we recognise as a functioning plant.

Part 1: What is growth?

At our present state of knowledge growth is best defined by describing the changes which occur as it progresses. Other disciplines, for instance mathematics, anatomy, biochemistry, help us to do this. We cannot here cover these subjects in detail, but this section indicates the way they can be used in the study of growth.

1

How is growth measured?

If we say that a plant, or part of a plant, is growing faster than another we have said very little unless we define what we mean by growth and measure how much growth is occurring. *Size* is the most obvious characteristic which we notice changing and this is expressed in terms of length, area or volume.

The precise size of living organisms is often difficult to measure, owing to their complex shape. Change in size is normally associated with change in weight and we are familiar with the use of both of these quantities from measurement of our own growth. However, although a child's size may be roughly measured on a linear basis by his height, children of the same height often differ widely in weight and vice versa. Adult humans stop growing in height but all too often increase in girth! A plant is even more difficult to measure linearly, owing to branching, to subterranean organs such as roots and tubers, and so on, which give it an extremely complex shape. Weight would seem to be a simpler measure of growth, but total weight including water content can also be very misleading. At the end of a hot day a plant's total weight, its fresh weight, may be less than it was at dawn for obvious reasons, although its dry weight, that is its fresh weight minus the weight of the water, will have increased owing to growth during the day. However, even dry weight can be misleading. The substances we include when we weigh a dried plant are proteins, fats, carbohydrates, pigments, salts and so on. If a plant in a salt marsh merely absorbs salt during the period of measurement, thus increasing its dry weight, it obviously cannot be considered to have grown. On the other hand, when a germinating seedling uses up, by respiration, some of its stored food reserve, while the root and shoot elongate, common sense tells us that the seedling is growing, even though it is decreasing in dry weight. Despite cases of this sort, dry weight is often used as a measure of growth, length is often used for simple organs like branches, stems and roots, and area is used for leaves.

Change in the quantity of protoplasm is often considered to be a more fundamental expression of growth. This is a difficult if not impossible thing to measure, but total protein is a good approximation to it since protoplasm, apart from its water, is largely protein. *Protein content* is often used, therefore, to investigate growth, but it is not easy to measure unless fairly complicated chemical apparatus is available. Protein contains the element nitrogen to about one-sixth of its total weight, and *total nitrogen* content is rather easier to measure than total protein, although it also presents difficulties. Moreover, not all the nitrogenous compounds in the plant are protein, and these will cause some errors.

Growth in size is almost always accompanied by increase in complexity, and the quantitative measurements described so far do not take this developmental aspect of growth into account. To measure this, research workers may count the *increase in numbers* of leaves, roots or other organs.

Differentiation is one of the most difficult growth processes to quantify. It is often observed at the microscopic level, and changing numbers of different types of cell can be counted.

Biochemical differentiation, namely the changing content of chemical constituents, enzymes or biochemical processes in cells and tissues, is now being intensively investigated. It can often be detected very early during growth, before any histological or anatomical changes, so that chemical or biochemical methods of measuring enzyme activity, quantity of nucleic acids and other constituents are now being increasingly used to estimate growth activity in its very earliest stages.

Usually more than one of these quantities is recorded to develop a picture of the amount of growth and development which is occurring in a plant or an organ. Figure 1.1 gives some idea of the measurable changes which take place in a particular case.

To summarise, growth cannot be universally measured, or even described, by the change of any single quantity. Changes in size, weight, complexity of form, types of organs and cells, complexity

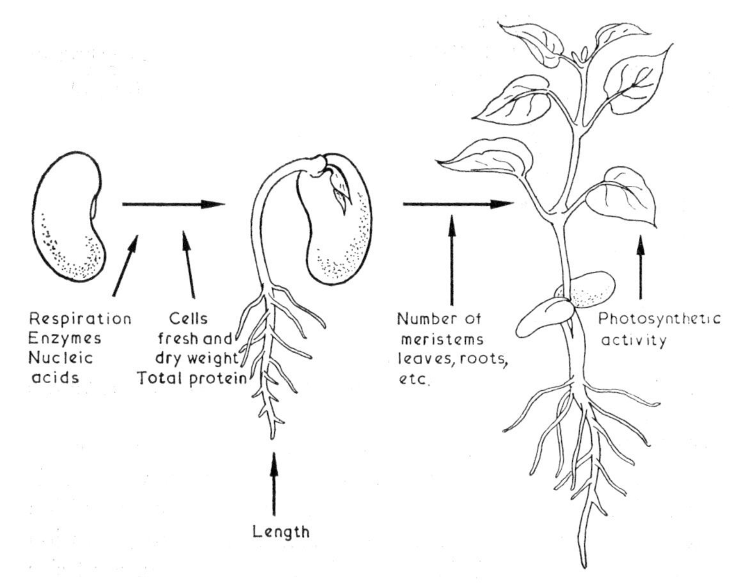

Fig. 1.1. Some of the many characters which change rapidly as a seedling grows. The stage indicated for each is the one where the most striking change may take place, but all the characters continue to change throughout growth

of chemical constituents and biochemical processes, can all be used to measure some aspect or other of the growth process, and these together can make an overall picture.

How do these measurable quantities change during growth? Figure 1.2 shows the time course of three of the simpler ones, namely the height or weight of a plant, and the area of a single leaf. All the curves are similar. Basically, they are S-shaped or sigmoid, and show that increase in size starts slowly, becomes more and more rapid, and then slows and stops. These phases are shown more obviously when the total increase during each successive equal time interval (which may be minutes, days or weeks according to the magnitude and duration of the growth rate) is plotted against time – see figure 1.3.

If we consider the simplest types of growing cell colony, e.g. a clump of bacterial cells with adequate nutrient supply during the

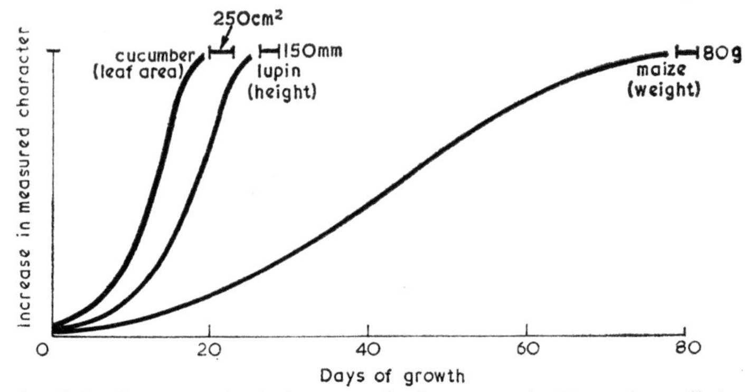

Fig. 1.2. Some typical changes during growth. Note that all the curves are S-shaped

period of measurement, it is obvious that the increase in size of the colony will be 'faster-faster', as the cell number will double when all the cells divide, double again during the next division and so on. Assuming the interval between division, the generation time, to be equal, and that the cells expand to a constant size before they divide we can express the growth of the colony in exact, and very simple, mathematical terms:

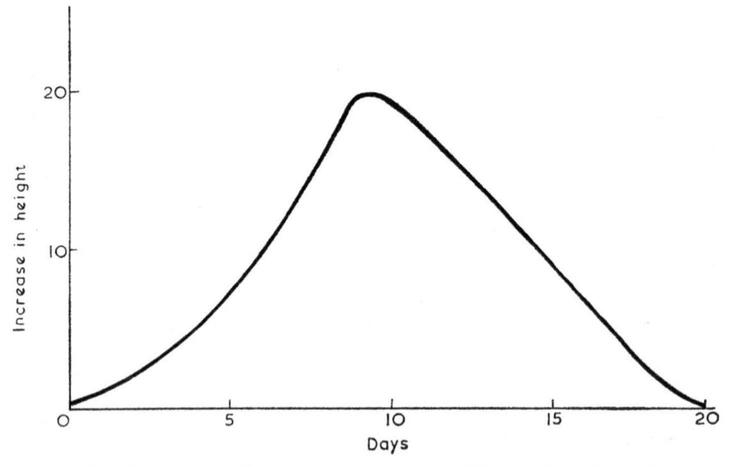

Fig. 1.3. Growth in height of lupin seedling. The data shown in Fig. 1.2 plotted as daily increments. (after Fogg)

Let N be the initial number of cells
N_t be the number of cells after time t
x be the number of generations during time t

Then the number of cells (which is in this case a measure of the size of the colony)

after one division will be $N \times 2$
after two divisions will be $N \times 2 \times 2$
after three divisions will be $N \times 2 \times 2 \times 2 = N \times 2^3$
after x divisions will be $N \times 2^x$

The graph of such a situation is shown in figure 1.4 and reminds us of the rising part of the growth curves in figure 1.2. This graph is an exponential curve, and the colony can be said to be growing exponentially.

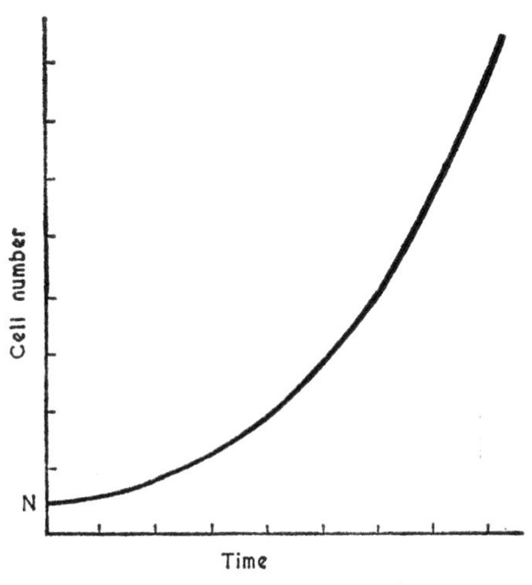

Fig. 1.4. Increase in size of bacterial cell colony with time

We can test the exponential nature of this curve very simply:

If $$N_t = N \times 2^x$$

then $$\log N_t = \log N + x \log 2$$

log 2 is a constant, so this equation becomes

$$\log N_t = \log N + \text{constant}.\ x$$

which is the equation for a straight line, usually symbolised by the general equation $y = mx + c$; in our case

$$v = \log N_t$$
$$m = \log 2$$
$$c = \log N$$
$$x = \text{number of generations}$$

As the number of generations is difficult to observe, but time intervals are, of course, easy to measure, it is convenient to express x in terms of time:

$$x = \frac{\text{time } t}{\text{generation time } g}$$

therefore, $\log N_t = \log N + \text{constant} \times \frac{t}{g}$

As g is also a constant (for the colony in question)

$$\log N_t = \log N + \left(\text{constant}_1 \times \frac{1}{\text{constant}_2}\right) t$$

and putting the term in brackets equal to a new constant k, we have

$$\log N_t = \log N + kt$$

which is also a straight line (figure 1.5).

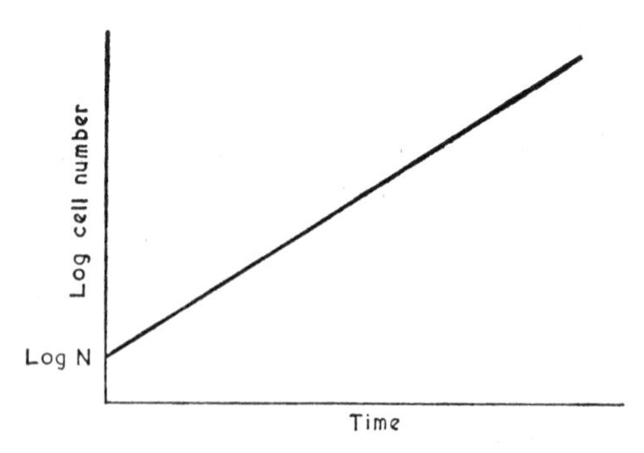

Fig. 1.5. A straight line is obtained if the logarithm of the cell number (from Fig. 1.4) is plotted against time

So if we plot the log of the cell number (or colony size) against the only other variable, time, and get a straight line, the growth is exponential.

As an example, let us start with 100 cells, cultured in a nutrient medium which is sufficient for optimal growth for at least 2 hours, all dividing at 20-minute intervals, the cells expanding to full size between divisions:

Time (*min*)	*Size of colony* (No. of cells)	*log cell number*
0	100	2·000
20	200	2·301
40	400	2·602
60	800	2·903
80	1600	3·204
100	3200	3·505
120	6400	3·806

$$\log N_t = \log N + kt$$
$$= 2 + kt$$

If the reader plots the cell number, and the log of the cell number, each against time, he will see the validity of the above argument.

The analysis we have just described is exactly the same as the compound interest law for invested money, where the increase (interest) at the end of each time interval (e.g. each year) itself grows by the addition of interest in subsequent time intervals. In both cases the *relative growth rate* (i.e. the interest rate) remains constant at all times, but the *absolute growth* increases exponentially, as the 'capital sum' (or number of cells) to which the constant relative growth rate applies, increases as time proceeds.

Observed growth curves of higher plants never exactly follow this mathematically perfect model for a variety of reasons:

(1) In real situations the nutrient supply becomes progressively depleted, and usually eventually exhausted, at some time during growth, thus becoming a limiting factor in the process. In plants, the ratio of the food-supplying organs, the leaves, to the rest of the plant (stems, flowers, roots) falls as the size increases and

thus the food supply does not keep up with the initial growth rate. Under these circumstances, we can expect the relative growth rate to fall more and more, eventually becoming zero when the leaves die. The absolute growth therefore also slows down and ultimately the organism will stop increasing in size. This situation is shown in figure 1.6, and curves *A* and *B* can be seen to resemble those of increase in size in figure 1.2 and in relative growth rate in figure 1.3.

(2) Not all the daughter cells are themselves able to divide, most of them differentiating into specialised cells and ceasing growth, even though the plant as a whole is growing. As a plant

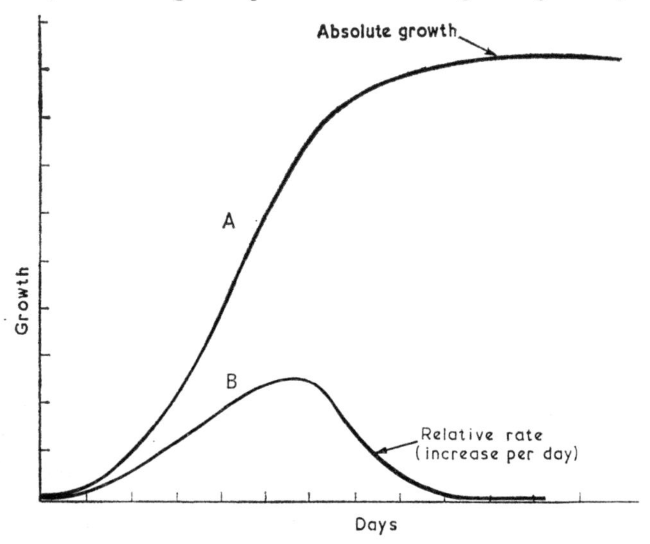

Fig. 1.6. Effect of progressive restriction of food supply on growth. The situation shown here is a theoretical one, but note that curves A and B resemble real results shown in Fig. 1.2 and Fig. 1.3 respectively

gets larger the number of such cells which are non-contributory to growth also gets relatively larger, and as we will see later, only in very restricted regions, the meristems and cambia, do cells continue to divide.

(3) The overall growth in size is made up of cell division

followed by cell expansion but different types of cell expand to very different extents, so no general rule can be formulated for this contribution to the whole growth.

(4) As we shall see later, various external and internal factors affect growth in complicated ways which are not directly related to the nutrient supply. Thus seedlings germinated in the dark will grow taller than those in the light even though all other conditions are identical.

(5) There may be an innate maximum size, that is a maximum cell number, which is imposed genetically. This is seen most obviously in animals, but although plants usually have an indeterminate growth, that is some parts of the plant can be induced to grow as long as suitable conditions are provided, individual organs such as leaves and flowers may have determinate growth, that is a predetermined maximum number of cells.

Summary

We can say that the growth curves of organisms and of organs, however growth may be measured, are usually sigmoid, showing an initial small increase or *lag phase*, a faster-faster or *log phase*, and then a plateau where growth has stopped, i.e. the growth rate has become zero. We can note here that at some time along this plateau degeneration or senescence sets in, leading to death. The part of the curve before the plateau is usually logarithmic to some extent but it rarely follows a perfect logarithmic course because of the complexity of factors, both internal and external, which influence and control the growth process.

2

Where does growth occur?

This section is not intended to be an anatomical description of plant tissue. This can be found elsewhere. However, it is obvious that anatomy and the development of structure are topics which cannot be separated easily.

The cells of which all plants are made vary greatly in size and structure. Cells of similar type often occur together to give tissues, although some cells may be scattered singly, or in small groups, through parts of the plant. It is important to realise that all cells differentiate to give their final form, and that apparently simple cells such as those which comprise parenchyma are as fully differentiated for their function as are, say, stomatal guard cells which have a more complex structure.

Growth in size of a plant or any of its organs takes place by the expansion of new cells derived by cell division. The two processes of division and expansion are usually separated in both time and space, but there is usually an obvious gradation from the one to the other: in higher plants the formation of new cells by cell division takes place mainly in regions called meristems.

Apical meristems

There is a meristem at the apex of every root and shoot, and this is capable of continued cell division to give indeterminate growth in length. We will see later that certain factors may suppress this capacity. The apex of a shoot is normally enclosed by young leaves which originate from primordia; these are groups of cells a short distance behind the apical meristem, which are also capable of division but do not go on doing so indefinitely. Leaves, therefore, have determinate growth. Under suitable conditions,

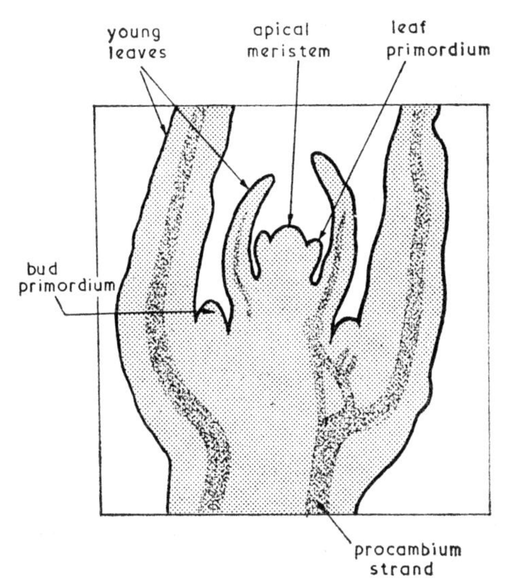

Fig. 2.1. Diagram of a median longitudinal section of a lilac stem apex. (after Robbins, Weier and Stocking)

flower primordia develop instead of leaf primordia. Lateral branches are derived from meristems which develop in the axils of the leaves. Sometimes these lateral branches are prevented from developing further than buds by factors which we will consider later. Figure 2.1 shows a typical leafy shoot apex.

Root apices do not have leaf primordia; they are protected by a root cap (figure 2.2). Lateral roots, unlike lateral shoots, are initiated in the deeper layers of the main root; cell divisions in the pericycle form a primordium, and the newly growing lateral root, with its own apical meristem, pushes through the cortex of the main root and erupts from the surface (figure 2.3).

Some plants have intercalary meristems which are not at the apices; grasses, for instance, have such meristems at the bases of the internodes, and they contribute to the growth of the stem.

Apical meristems cause much of the obvious growth changes in plants and their activity is said to give rise to primary growth.

Cells which are cut off behind the meristems are the ones which expand – those forward of this region continue to be meriste-

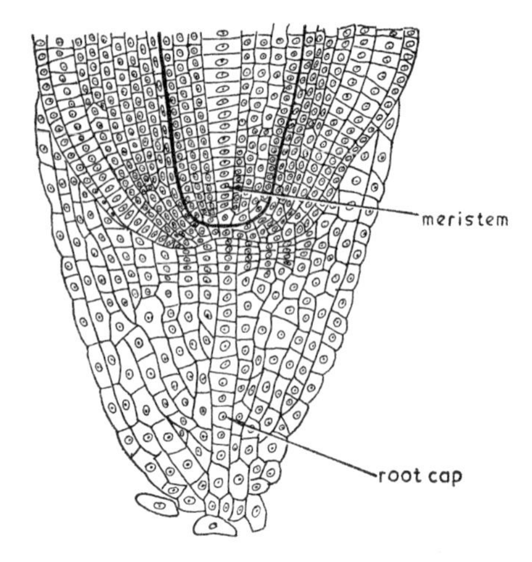

Fig. 2.2. A typical root apex. (after Robbins, Weier and Stocking)

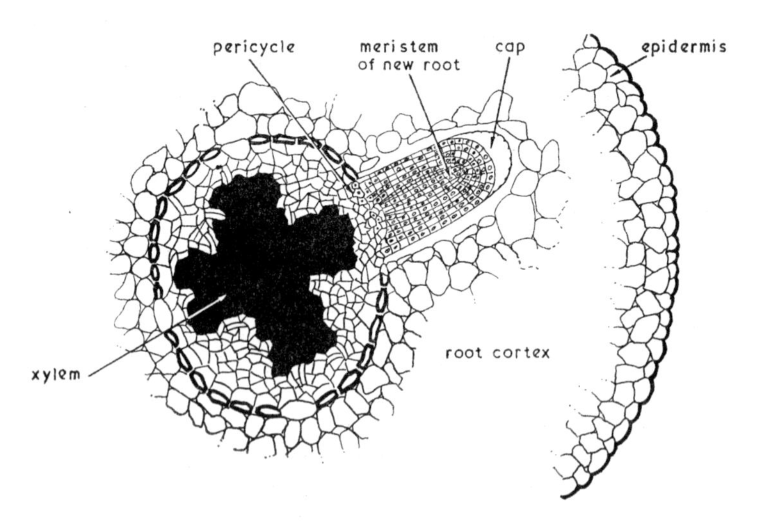

Fig. 2.3. Origin of a lateral root. (after Telfer and Kennedy)

matic. The expanding cells change not only in size, but also in structure, that is they differentiate to form the many different tissues which are found in the mature organ. Figure 2.4 shows diagrammatically the growth and differentiation of a root and the changes which occur in single files of cells.

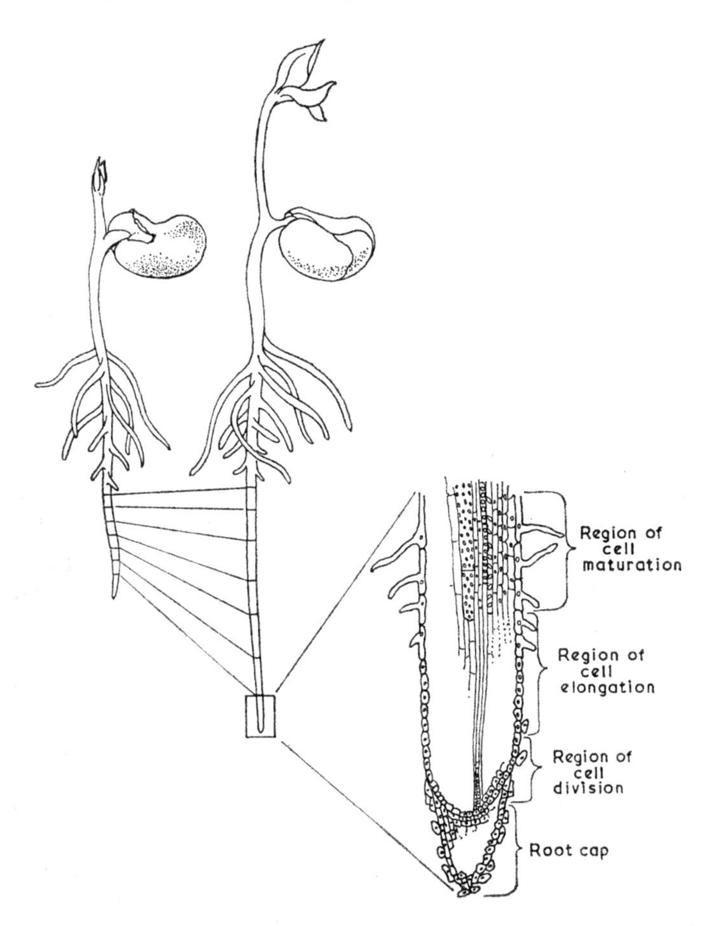

Fig. 2.4. Changes along a root during growth. The cells can be considered to be older the further away they are from the tip. (after Telfer and Kennedy)

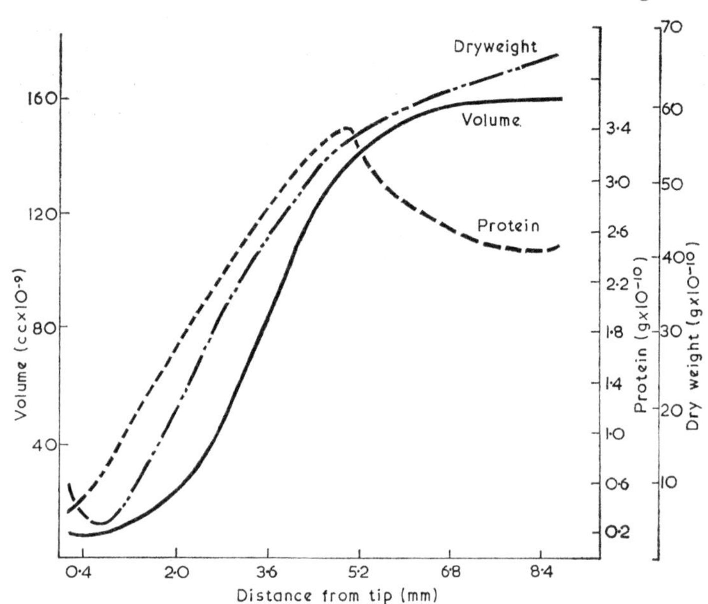

Fig. 2.5. Changes in volume, dry weight and protein content per cell along the axis of a pea root. Cells are older the further away they are from the tip. (after Brown, Reith and Robinson)

Because the meristem cuts off cells behind it, it is evident that the further back a cell is situated the older in the developmental sequence it must be. The situation shown in figure 2.4, therefore, represents a linear demonstration along the axis of the root of the changes in cells as they grow. It is fairly easy to cut the root into cylinders and to count the number of cells in each. This is done by macerating the tissue, i.e. treating it with strong acid, which causes the cells to separate from one another to give a suspension in liquid; the number of cells in a measured drop of this can be counted under the microscope. The fresh weight, dry weight and protein content of the cylinders, divided by the number of cells, gives these quantities *per average cell*. Figure 2.5 shows some of the changing characteristics along a root axis; these can be taken to be changes with time. Later we will consider the factors which influence these changes in newly formed cells.

Cambia

The vascular cambium is another kind of meristem. Cambial cells differ from those of apical meristems in position and shape. Instead of being approximately isodiametric (of equal length along all diameters) they are flattened and may be much longer than wide. Cambial activity is important in the secondary growth of stems and roots, where it causes increase in girth (figure 2.6).

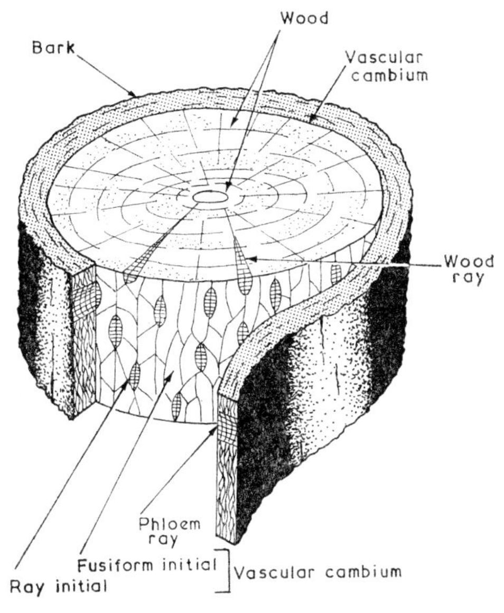

Fig. 2.6. Diagram of location of cambium causing increase in girth of a woody stem. (after Robbins, Weier and Stocking)

Elongated cambial cells give rise to sieve tubes or vessels; the parenchyma of the rays is formed by smaller cells.

Cambial cells divide along a plane parallel to their flat surfaces, and like other meristematic cells they form young unexpanded daughter cells which subsequently enlarge and differentiate (figure 2.7).

Cambia are also found just beneath the surface of organs such as potato tubers, and woody shoots, where they give rise to the

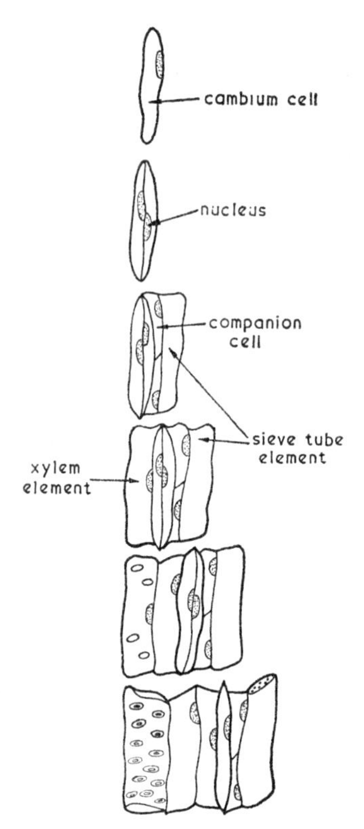

Fig. 2.7. Diagram of division and differentiation of cells arising from vascular cambium. (after Robbins, Weier and Stocking)

phellem or corky peel. Such a cambium is called a *phellogen*. A phellogen can be induced to form below the wounded surface when a potato tuber is cut. The parenchyma cells near the surface, which would never have divided in the unwounded tuber, start to divide after a few days. They give rise to a layer of cells which first resembles a cambium, and which then forms a surface phellem, healing the wound. Under certain conditions of hormone and nutrient treatment, however, areas of this dividing layer can be induced to form large groups of meristematic cells giving indeterminate growth; the result is an enlarging mass of parenchymatous tissue called a *callus*. The wounded surfaces of pieces cut from plant tissues can in this way be the sources of growing masses of tissue known as *tissue cultures*. In many cases, further treatment with hormones can cause cells derived from these 'meristems' to differentiate to give recognisable tissues (phloem, xylem) and even organs such as roots and shoots. Eventually a whole new plant may be formed (Plate 13).

It is evident, therefore, that fully differentiated parenchyma cells can become converted to a juvenile condition, an occurrence sometimes called de-differentiation, in which state they can divide and give rise to new and different tissues. Investigation of tissue cultures has become an important means of discovering the factors which cause cells to divide and differentiate. Any information about the changes which occur, or which can be caused to occur, in these cultures can help to explain the causative agents and control processes in normal meristems and their associated zones of differentiation.

Diffuse growth

Not all cell divisions occurring during growth are confined to the meristems. Dividing cells can be found in the expanding zone behind the meristems, or scattered, for instance, in expanding leaves; stomata are formed by a special sequence of divisions in the epidermis. Such divisions do not continue indefinitely, but die out as the tissues mature.

Summary

The increase in numbers of cells occurs largely in the meristems, of which there are two main types: apical meristems and cambia. Increase in size is brought about by expansion of daughter cells derived from cell divisions in the meristems.

Mature cells can be induced to de-differentiate, i.e. regain the capacity for cell division, when pieces of tissue are cut from the parent plant. These new divisions give rise to a mass of growing callus called a tissue culture. These calluses can differentiate into organs or even whole new plants. They provide us with a means of investigating the factors which cause cell division, expansion and differentiation.

3

What happens when a cell grows?

We have already seen that the two basic cellular changes which lead to growth in size are division and expansion. Differentiation occurs during and subsequent to the latter process.

Cell division

Much is known about the detailed cytological changes, especially nuclear behaviour, during cell division and it is unnecessary to go into them in this book (see Torrey). Little is known about the causes of these events. For instance, in meristematic cells the nucleus may be triggered to divide when the cell reaches a certain size, but how the nucleus detects this critical volume is unknown. The dividing nucleus in turn, triggers events leading to cytokinesis, that is division of the whole cell (see Plate 1). Profound biochemical changes take place in the nucleus and in the rest of the cell at all stages of division, including the apparently quiescent interphase when DNA doubles in the nucleus in preparation for subsequent mitosis. RNA synthesis occurs mainly during prophase and telophase, and is located in the nucleoli as well as in the cytoplasm. Very little work has been done on the changes in individual enzymes in the nucleus or for that matter in the cell as a whole, because of the difficulties in detecting them in a dividing cell surrounded by cells which are at different stages of division, or not dividing at all; but we can guess that an elaborate timing sequence is imposed upon their synthesis also.

Each step in the mitotic sequence must in some way trigger the next, but 'on top of this self-regulating machinery there are imposed commands from the body of the organism as a whole which control the cell cycles [divisions] so that they react to the genetical and environmental demands in a predictable fashion' (Clowes and Juniper, *Plant Cells*, 1968).

Water makes up the bulk of all cells, accounting for some 80–90% of the total weight and volume (cells do not contain gas spaces). In mature plant cells most of this water is contained in the large central vacuole. Meristematic cells have many small vacuoles scattered in the cytoplasm, but as the daughter cells expand, these small initial vacuoles enlarge and fuse together, eventually forming a single vacuole which becomes the largest part of the cell, and this goes on expanding during growth. The increase in size and fresh weight of a cell, therefore, is mostly accounted for by increase in water content, and uptake of water must be an important growth process (Plate 3).

Plant cells absorb water mainly by osmosis* and we can show by experiment that if an expanding plant tissue, for instance a root tip, is immersed in a solution which has a higher osmotic potential than its own vacuolar liquid (sap) growth in size ceases. We would learn from this experiment that the other major manifestation of growth, differentiation, is also prevented, showing that increase in size and increase in complexity are closely linked processes (Plate 3).

During expansion it is found that the osmotic potential of the sap does not change very much, if at all, implying that an increase in the amount of soluble matter (sugars, salts, amino acids, etc.) in the vacuole of each cell keeps pace with the increase in volume. As the cells need to synthesise or absorb nutrients for this purpose, this aspect of growth cannot be considered to be merely passive. These nutrients come partly by translocation from surrounding tissues, and partly by breakdown of food reserves (e.g. starch) in the cell itself. Nutrients are also required for increase in weight and complexity of the cell wall and of the protoplasm.

We will see later that synthesis of the complex polysaccharides and proteins which help make up these parts of the cell requires energy which is derived from respiration, so some of the incoming nutrients are used as respiratory substrates. Rapidly expanding tissues can be expected to have a higher respiratory rate than do mature tissues, as they need more energy for their extra growth

* See Baron, W. M. M. (1966) *Water and Plant Life* (in this series).

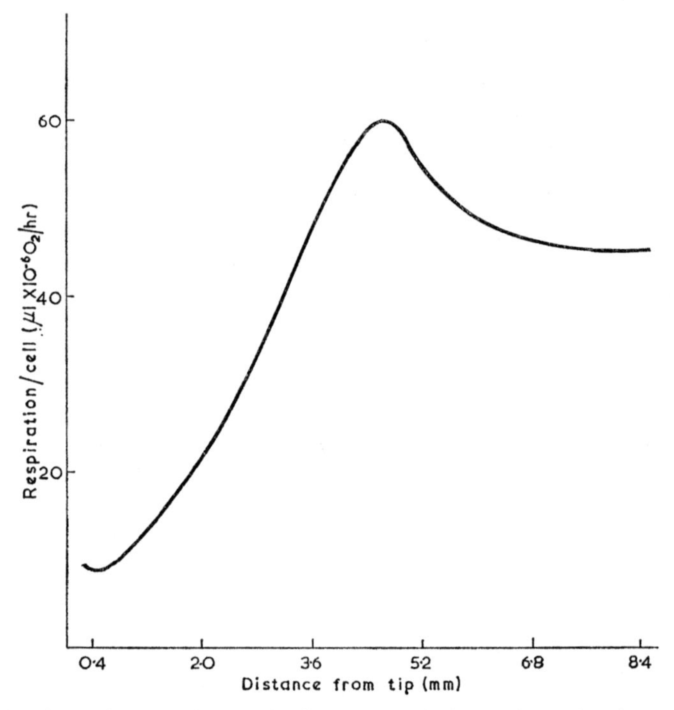

Fig. 3.1. Change in respiration per cell along the axis of a pea root. The respiratory rate rises rapidly as the cells expand. (after Brown, Reith and Robinson)

activities over and above that used for other cellular processes. Investigation of a growing root show this to be so (figure 3.1).

Before we see why energy is necessary for the synthesis of complex compounds in the cell, we will consider what changes occur in the structural components, and for this we need to know something more about the cellular structure itself.

The cell wall

When a cell divides, the cell plate is laid down by vesicles which aggregate at the equator of the mother cell after the nucleus has divided (Plate 1). The middle lamella is first formed, and

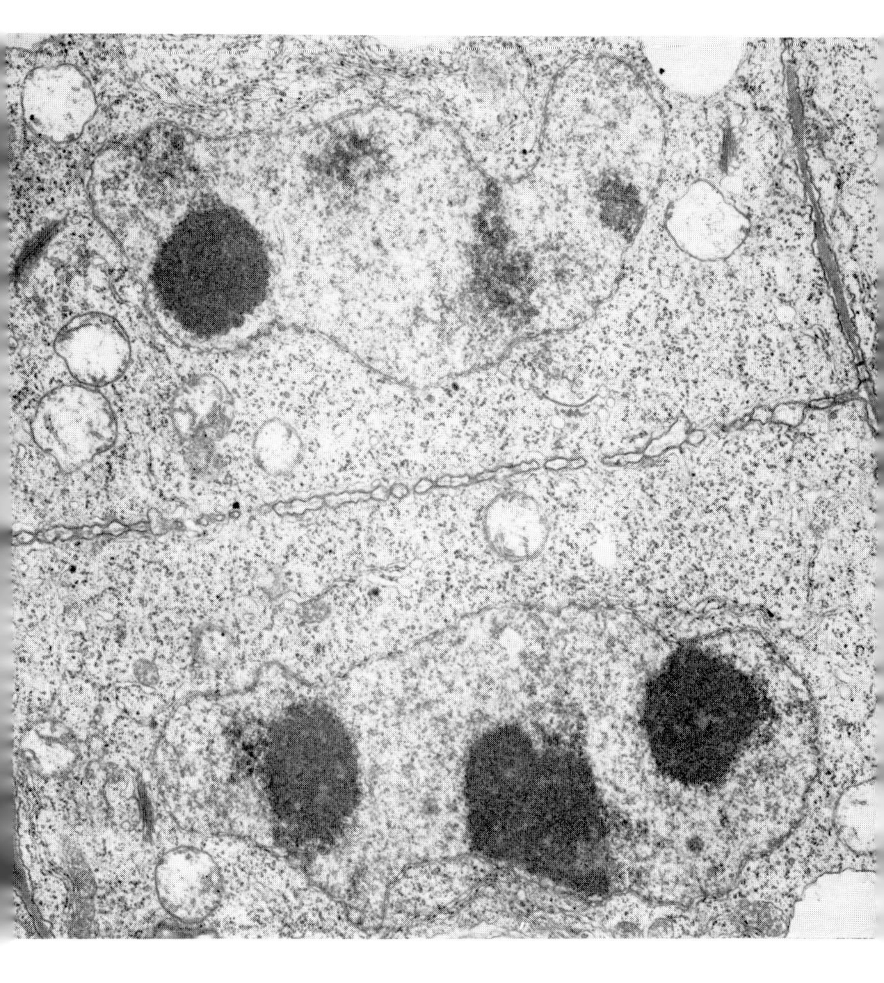

Plate 1 *An electron micrograph of a very early stage in cell wall formation in bean cells.*
A cell of the hypocotyl has just divided and the daughter nuclei can be seen clearly. The cell plate, running transversely across the photograph, is being formed from a layer of vesicles. Magnification ×11,400.

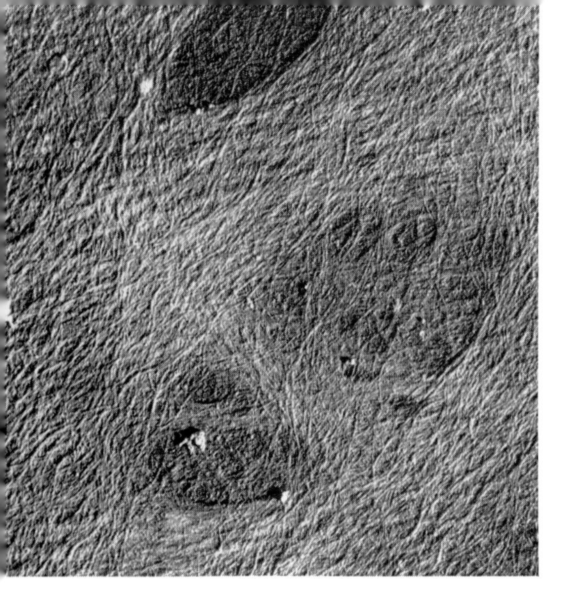

Plate 2 (*a*) *Electron micrograph of the primary wall of an artichoke cell.*
This is a surface view after the soluble polysaccharides of the wall have been removed leaving the cellulose microfibrils. Pit fields can be seen clearly. Magnification ×9,100.

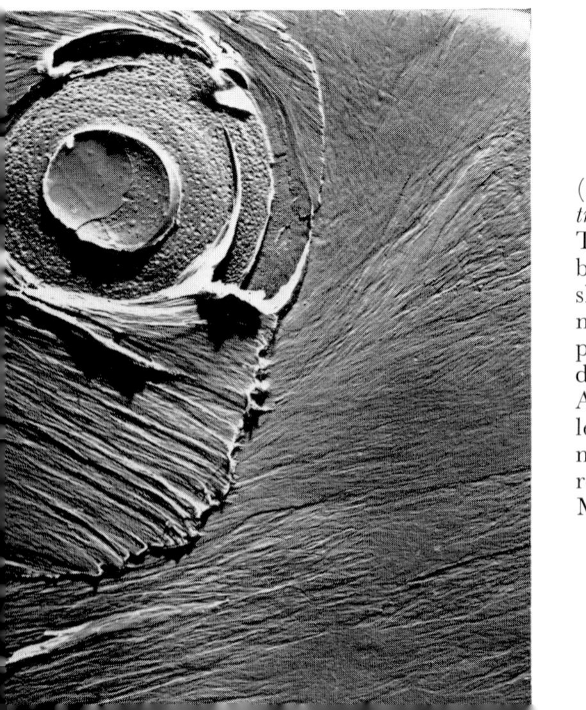

(*b*) *Secondary wall of pine tracheid.*
The outer layer, left, has been partly torn away to show an inner layer. The microfibrils are more or less parallel, but run in different directions in the two layers. A bordered pit is seen top left. (This electron micrograph is of a carbon replica of the specimen). Magnification ×3,500.

this consists largely of pectin, a polysaccharide. The primary wall is rapidly laid down against the middle lamella, and this contains cellulose fibrils embedded in pectic substances and other polysaccharides called hemicelluloses. This wall is very thin, 1–3 μ thick, and it stretches readily as the cell expands. This stretching is plastic (irreversible, like dough) rather than elastic (reversible, like rubber), that is, if the increased water content causing the stretching is removed the wall would not return to its original shape. The primary wall can be seen under the electron microscope as an apparently random network of cellulose fibrils (Plate 2(*a*)), the matrix of pectins and hemicelluloses being difficult to see, as they are amorphous. It is important to remember that the electron microscope can be used only with rigorously dried specimens and shows a collapsed impression; in the living cell the wall is 'filled out' with water which is held fast by attraction to the polysaccharides, particularly the hemicellulose molecules.

The chemical composition of cell walls obviously cannot be determined unless they are separated from the protoplasm. When this is done by grinding up the tissue and washing away the protoplasm with various reagents, all walls, including primary walls, are found to contain small amounts of fats and protein as well as carbohydrate. Until recently these were thought to be bits of the protoplasm, possibly plasmadesmata, stuck to the wall, as 'pure' walls are difficult to separate from the rest of the cell contents, but now it is thought that they are an integral part of the wall. Their function is still unknown.

Expansion growth of a cell may take a few hours or days and while it occurs considerable amounts of cellulose and hemicellulose are added to the wall, eventually making it much more rigid. This development of the secondary wall may continue for a considerable time after expansion growth has ceased, and eventually most of the wall of the mature cell consists of this secondary material. It confers on the cell its final shape and much of its mechanical strength (the rest comes from support from surrounding cells and, in living cells, from osmotic turgor). At this stage the wall is elastic rather than plastic.

Differential rate of wall deposition must account for local thickening within a cell, for instance, in collenchyma, in xylem and in stomatal guard cells. Plate 2(*b*) shows some aspects of secondary wall formation; it is important to remember that, although it is convenient to distinguish between primary and secondary walls, wall growth is a continuous process and that there is not normally a sharp transition in time from one stage to the next.

Many specialised compounds can be found in the walls of differentiated cells, depending upon their structure and function. These include notably lignin in woody tissues, suberin in corky layers, cutin in cuticular surfaces of epidermal cells and so on. More information about these can be found in anatomy texts.

Wall formation

For some time cellular physiologists have speculated about how the cell wall is made by the cytoplasm. The incorporation of high molecular weight, and especially insoluble, compounds is difficult to visualise, particularly if the wall is considered as a non-living envelope with no metabolic capacity of its own.

Recently it has been shown that the golgi body is involved. This organelle is very prominent in growing cells. It gives rise to vesicles which contain cell-wall polysaccharides, apparently hemicelluloses. These vesicles move towards the plasmalemma and discharge their contents into the wall by reversal of the process called pinocytosis (figure 3.2). The cellulose fibrils appear to be extruded into the wall by small bodies attached to its inner surface, but the means by which these fibrils are made and positioned is still obscure.

Metabolic and synthetic activities of the wall itself are now being suggested, at least for growing cells, and the view that it is merely an inert framework is becoming modified. The presence of proteins in the wall supports these ideas, especially as these proteins change markedly during growth. Investigations of these changes have only recently started, partly because it was not originally thought that wall contained protein, for the reason given earlier, and partly because the techniques of detecting and investigating such small quantities of protein as occur (usually less than 1% of the dry weight of the walls and often less than 0·1%)

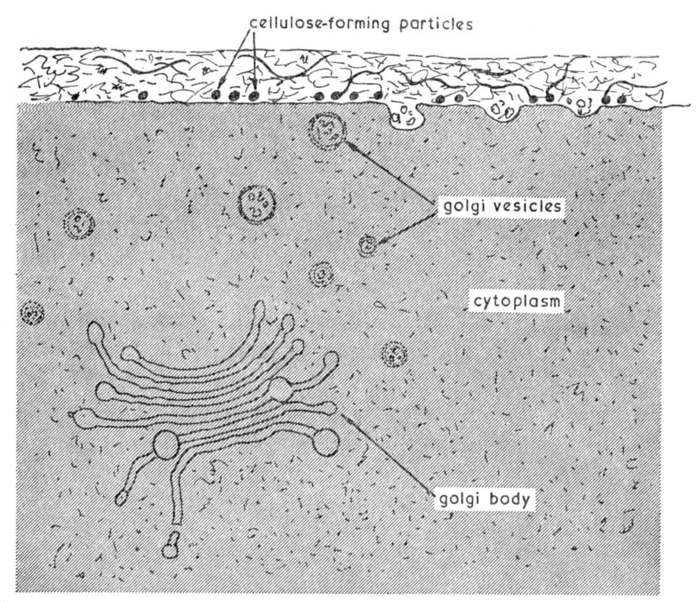

Fig. 3.2. Diagram of cell wall growth. Vesicles are shed from the golgi body and move through the cytoplasm towards the wall. They discharge hemicelluloses and pectins through the plasmalemma and into the wall by reverse pinocytosis: an imaginary sequence of this process is shown top right. Cellulose fibrils are synthesised by small bodies lying at the interface of the wall and the plasmalemma

were not, and indeed are still not, very good. Two observations using tissue cultures as the experimental material have shown that the cell-wall protein may be important in both wall structure and growth. First, it was found that the protein has a very high content of hydroxyproline, an amino acid not found in protoplasmic proteins and which had previously been thought to be restricted to the structural proteins (collagen) of animals. This hydroxyproline-rich protein of walls was found to increase shortly after cutting out pieces of tissue to start the culture, that is when growth is just beginning.

Secondly, the cell walls of tissue cultures starting to grow contain certain enzyme activities, and some of these increase during the early stages of growth. Some of them, for instance, invertase, are confined to the cell wall, being absent from the cytoplasm; it also seems significant that this enzyme is associated with the elongating zones of growing organs, e.g. just behind root meristems.

We can speculate that the protein which contains hydroxyproline is involved in some way in the structure of the wall, a new idea as the structural properties of the wall have hitherto been ascribed solely to the polysaccharides. As the amount of protein is so small, it cannot be important as mere bulk, but would have to have a specialised function, possibly providing a framework on which the wall polysaccharides are constructed.

The wall enzymes, on the other hand, may be expected to play a part in metabolism, and thus presumably in wall growth. Both of these aspects are interesting new fields for research which are being actively pursued.

The protoplasm

During the growth of a cell, great changes must occur in the protoplasm as well as in the wall. Indeed, the differentiation which is shown by the wall in such diverse cell types as sieve tubes, guard cells, tracheids and so on, must arise by the controlling influence of similarly differentiating cellular protoplasm. We know only a little about this as protoplasm is a highly complex material whose structure is by no means fully elucidated yet, but some at least of its basic properties might be expected to change during growth.

Enzyme changes during expansion and differentiation

The protoplasm of a cell contains several hundred, possibly thousand, different enzymes which drive the cellular processes. As these processes change with growth, so do the enzymes. Changes in some enzymes in root cells have been followed using the serial section technique described earlier, and figure 3.3 shows that different enzyme activities change differently, which is what we might expect, as the cells mature. These changes can be mea-

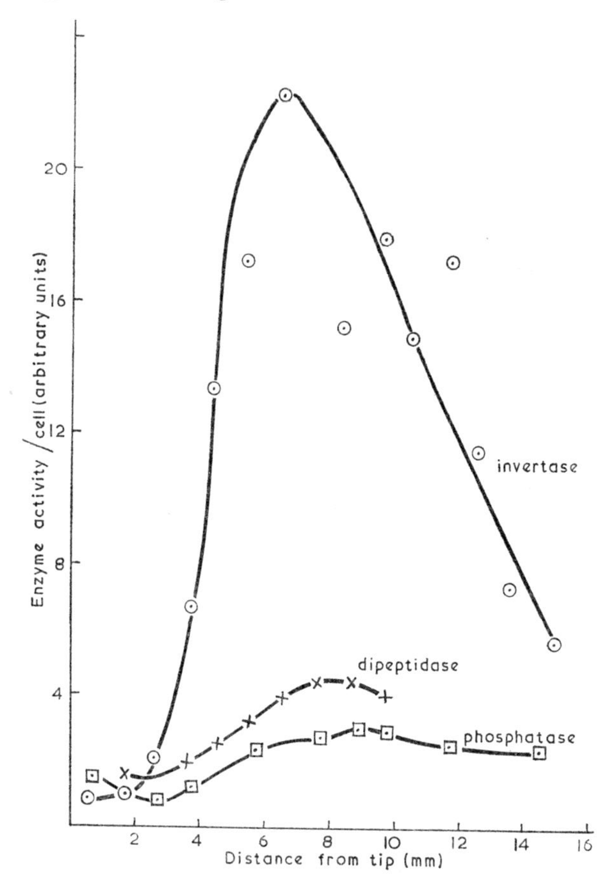

Fig. 3.3. Changes in some enzyme activities per cell along the axis of a pea root. Cells are older the further they are away from the tip. (after Brown, Reith and Robinson)

sured in a rather different way, namely as the amount of an individual enzyme activity per unit weight of protein. Again, as we would expect, figure 3.4 shows that some enzymes increase in proportion to the total protein, and some decrease.

We do not know why particular enzymes change in this way

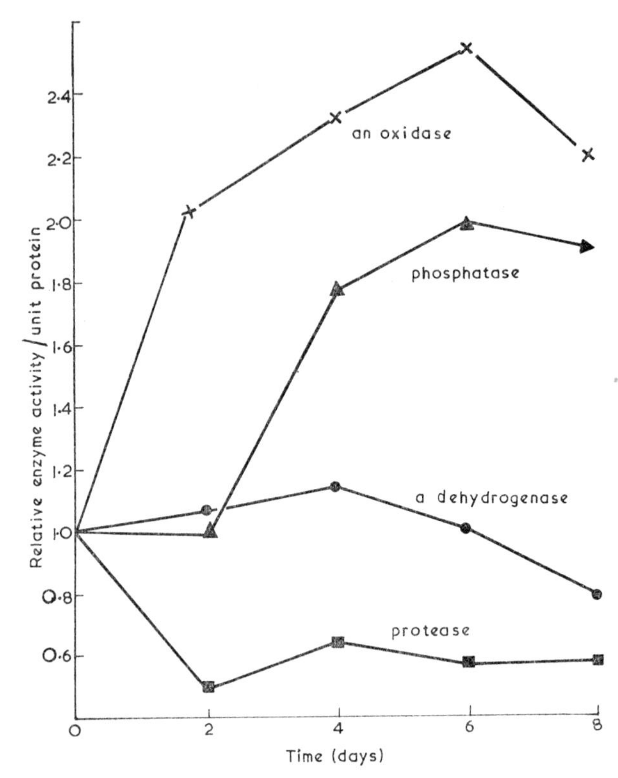

Fig. 3.4. Changes in relative amounts of some enzymes during growth of pea roots. Some enzymes increase and others decrease per unit amount of protein. (after Heyes)

as we do not yet know how all the individual ones are organised to work together, but from research along these lines we will be able to deduce which enzymes are important at different stages of growth and hope to discover eventually how they all fit together to perform the processes which make up growth. But protoplasm is not just a bag of enzymes; it contains many different types of organelles having specialised functions, each containing enzymes in a well-ordered structure.

Changes in the organelles; subcellular differentiation

Some of the more common organelles found in higher plant cells are shown in Plates 3(*a*) and (*b*). We would expect to find changes in them during growth and especially in those involved in energy production and biosynthesis for reasons described earlier. It is only in the last few years, since electron microscopes have become the stock-in-trade of the biologist, and since biochemists have elucidated biochemical functions of some cell organelles, that this aspect of growth has become an important field of research.

Mitochondria are the seats of respiration and, therefore, of the production of ATP, and it has recently been shown that the increase of respiration which is found when tissues begin to grow is correlated with an increase in the numbers of mitochondria in the cells. For this work tissue cultures have been used, as here the change from the non-growing to the newly growing state can be seen most markedly. Increase in respiration and mitochondrial numbers can be shown within a few hours (and in nucleic acids and some enzymes within even shorter times). Mitochondria can themselves divide independently of the nucleus, and this is probably how they increase in numbers. They contain DNA, the basic material needed for protein, and thus enzyme, synthesis (see below), and they can be considered to lead a somewhat 'independent' existence within the cell.

Perhaps the best-known example of subcellular differentiation is the case of the chloroplast. Young cells which are going to become green as they mature, contain proplastids which develop into chloroplasts through a sequence of changes which is easily observed by electron microscopy.

In angiosperms, cells which normally become green when exposed to light do not do so in the dark, i.e. in etiolated plants; here the synthesis of chlorophyll progresses only as far as its precursor protochlorophyll which is colourless, and at the same time the chloroplasts do not develop properly, suggesting that chlorophyll is not merely present in chloroplasts but is an integral part of their structure. The sequence of development is different in this case.

Exposure to light in the early stages of this restricted development leads to conversion of protochlorophyll to chlorophyll and to normal development of the chloroplast. So here we have an example of the effect of a single external factor, light, on biosynthesis and subcellular differentiation, resulting in an easily observed overall effect on the plant (Plate 4(*a*)–(*c*)).

Protein synthesis

All of these changes in the protoplasm involve protein synthesis, although many other biosynthetic activities (e.g. fat, carbohydrate synthesis) may be specific to particular processes.

'The control of protein synthesis means the control of growth.' As we are concerned in this book with the control of growth, we need now to consider protein synthesis in some detail, with the particular view of discovering how it can be manipulated during growth.

The mechanism of protein synthesis is now fairly well understood from work on micro-organisms and animal cells, and evidence from plant tissues indicates that the process is basically the same there too. We are not going to cover the biochemistry in detail, but will summarise the important features, and consider likely control points.

The protein make-up of the cell is fundamentally determined by the nucleus. The genes exert their influence on the development of the cell by passing instructions to the sites of protein synthesis in the cytoplasm. This is done in the form of 'messenger' molecules belonging to the class of compounds called ribonucleic acids (RNA). The genes themselves provide the information by virtue of their content of a similar class of compound, deoxyribonucleic acid (DNA), and the messenger RNA (mRNA) molecules are essentially a mobile copy (transcription) of the static genes. The mRNA molecules are 'read' (translation) by the protein synthetic system in the cytoplasm which then makes the corresponding proteins (enzymes, structural proteins). Enzymes formed in this way can then perform their parts of specific biochemical functions such as respiration, synthesis of cell-wall materials and so on. We can summarise the sequence of events as in figure 3.5.

Control may be exercised at any of the intermediate steps, the further back in the sequence it occurs, e.g. at the transcription of the information from the DNA to the messenger RNA, the more profound the results.

We can see microscopically that every diploid nucleus in a plant contains the full complement of chromosomes, whatever its cell type. We can also demonstrate that a starting tissue culture, consisting of only a few similar cells, can give rise to many different tissues, and even to a whole fully differentiated plant

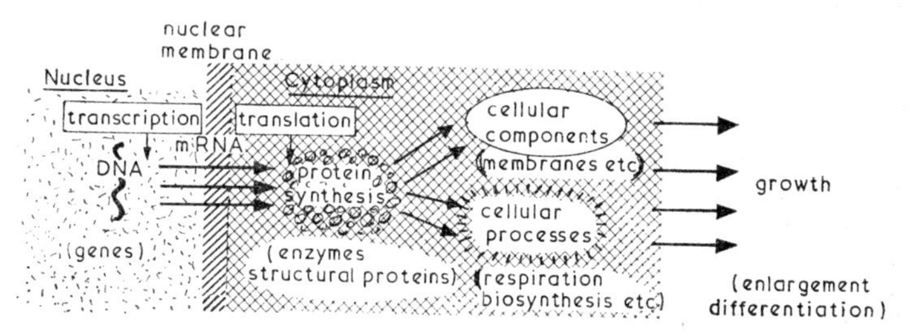

Fig. 3.5. Summary of transfer of information, which is required for growth, from the nucleus to the rest of the cell. Structural components of the cell, as well as agents (e.g. enzymes) which catalyse cellular processes must be synthesised in the cytoplasm

of its own species (Chapter 10). These pieces of evidence show that the information for differentiation into every cell type must be present, but latent, in all nuclei, and in the case of the tissue culture can be transmitted to new cells during repeated subsequent divisions.

We must infer from this that during the growth and differentiation of any particular cell, only the information relating to that cell is transcribed from its nucleus, information from all the other genes being suppressed: for instance the complete suppression, or 'switch-off' of all genes, except those which lead to changes towards collenchyma development will result in a young cell becoming a collenchyma cell – those genes can be considered to be the only ones which are 'switched on'.

There may be other control points later in the sequence, and if we know the intimate details of the process, we can speculate about where these controls may lie. This aspect of the subject is a very recent one. It falls into the realm now called molecular biology, but although the details are very complex, an outline of the sequence is easily understood.

Messenger RNA passes out of the nucleus, probably through pores which can be seen in the nuclear-bounding membrane (Plate 3(*a*)). It is possible that these pores increase in number when a cell grows and decrease when it matures. Each thread-like messenger molecule, which contains the information built into its complex chemical structure, associates with several (about six) of the tiny particles called ribosomes (Plate 3(*a*)). Ribosomes derive their name from their content of RNA, albeit a different type of RNA from that of the messenger molecules. The group of ribosomes attached to a messenger molecule is called a polyribosome, or polysome. While attached to the messenger, each ribosome is able to join amino-acid molecules together to make a protein molecule. The sequence of the different amino acids in the protein is determined by the message or *code* which the ribosome 'reads' from the mRNA molecule, and it does this while passing along the messenger from one end to the other. When the protein molecule is complete, i.e. when the ribosome has come to the end of the messenger, the protein, ribosome and messenger separate into the cytoplasm where the messenger is eventually destroyed. It appears therefore that a constant flow of messengers is needed for continued synthesis of a particular protein and that a messenger molecule can code for only one type of protein, as different proteins would require differently coded messengers.

Amino acids in the free state are not used by the polysomes: later we will see that they are made reactive, or activated, and become attached to yet another type of RNA, called transfer RNA as it transfers amino-acid molecules to the polysomes. The process is summarised in figure 3.6.

In addition to the three kinds of RNA involved in protein synthesis, namely mRNA (messenger), rRNA (ribosomal) and sRNA (transfer, the small letter 's' stands for soluble to show that it is not particulate) the reactions are catalysed by various enzymes and require certain co-factors. An important one of these latter is the high-energy compound ATP which is required for activation of the amino acids, and for synthesis of RNA itself. We know that the process as a whole is highly energy requiring,

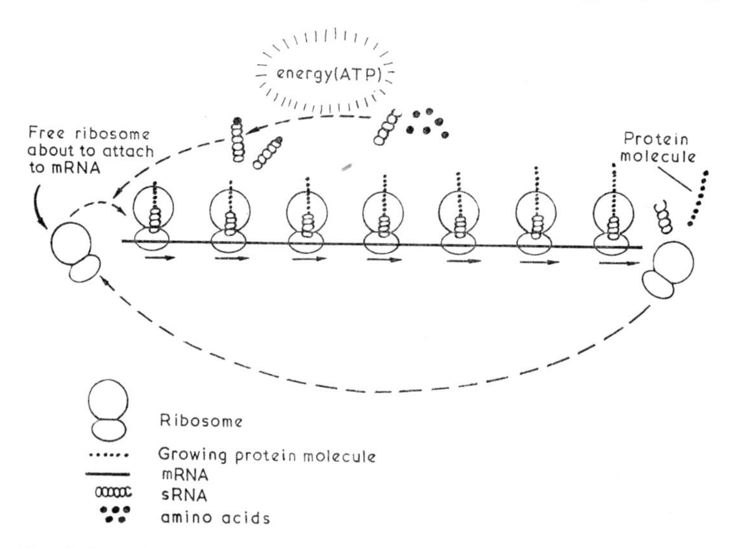

Fig. 3.6. Diagram of protein synthesis by a polysome (after Watson)

as anything which prevents respiration (inhibitors, lack of oxygen) stops protein synthesis.

We can infer, therefore, a high demand for ATP molecules – and not only for protein synthesis: in higher plants growth as a whole stops when respiration is prevented, and so do such other vital processes as translocation and salt uptake. Why is this?

The need for energy during growth

We can take, as examples of energy-requiring events, three biochemical processes which constitute part of growth (there are others, of course): (*a*) polysaccharide synthesis which must occur for cell-wall enlargement; (*b*) protein synthesis for protoplasmic increase; (*c*) nucleic-acid synthesis for nuclear division and increase in the protein synthetic system itself.

ATP is produced by photosynthesis and also by respiration. All cells respire. Even photosynthetic cells, when in darkness, must produce their ATP by respiration, and when the plant is in the light ATP is too unstable a substance to be translocated to non-photosynthetic tissue; photosynthetic energy is stored or translocated in the form of carbohydrate and released by respiration. As growing tissue, e.g. root tips, secondary growth, stem apices, is not normally photosynthetic we find that respiration is the major source of ATP.

Before we consider the specific requirement for ATP in our three processes, it is convenient here to summarise ATP production by respiration: further details can be found in any elementary biochemistry text.

The stored energy of glucose is used to make ATP molecules, and this process can be divided for convenience into two phases. In the first of these, glycolysis each glucose molecule ($C_6H_{12}O_6$) is transformed to two molecules of pyruvic acid ($C_3H_4O_3$). During the reactions which bring this about, two molecules of adenosine diphosphate (ADP) are combined with two inorganic phosphate groups ($H_2PO_3^-$, often denoted P_i) to form two molecules of ATP. ADP can be written A—P∼P (where A represents the base-sugar complex adenosine, and P a phosphate group) and ATP as A—P∼P∼P. The symbol — is a bond whose rupture (by hydrolysis) yields relatively little energy, and ∼ represents a 'high-energy bond' which yields about three times as much. It is the 'high-energy bond' which imposes instability, a readiness to react chemically, on the ATP molecules and, according to circumstances, the single terminal phosphate group, or a pair as pyrophosphate (P∼P, often denoted PP_i, actually $H_4P_2O_7$) may detach from triphosphates of this type.

The further degradation by the tricarboxylic acid cycle (or Krebs cycle), followed by oxidative phosphorylation, of each pair of pyruvic acid molecules in the cell to CO_2 and water leads to the production of a further 36 ATP molecules. At the same time six molecules of oxygen are absorbed for each pair of pyruvic acid molecules, i.e. for each glucose molecule originally entering the process. The familiar summary equation for respiration can then be written as:

$$C_6H_{12}O_6 + 6O_2 + 38ADP + 38P_i \rightarrow 6CO_2 + 6H_2O + 38ATP$$

The important fact to note here is that 38 inorganic phosphate groups are converted into the high-energy compound during the uptake of 12 atoms of oxygen. This gives a phosphate to oxygen ratio of about three, i.e. the P/O ratio = 3.

We may thus take it that each oxygen atom consumed can lead to the conversion of energy from the starting carbohydrates to three 'high-energy bonds'. If we have knowledge of the use of these bonds in a particular process, we can calculate the oxygen requirement of the process.

(*a*) ***Polysaccharide synthesis.*** Cell walls consist of several polysaccharides (see page 23) and we know in detail how some of them are made. In all known cases, monomers (such as glucose) are built into polymers (e.g. cellulose), and ATP-like substances are involved: these are the sugar nucleotides (figure 3.7). Guanosine diphosphate glucose, GDPglucose, appears to be the key intermediate in cellulose biosynthesis, and others like UDPxylose and UDPglucuronic-acid in the synthesis of hemicellulose.

We will take cellulose as a typical example. In the first step each glucose monomer, derived from food reserves, is converted to glucose phosphate. One of the reactions which can do this involves the transfer of a phosphate group from a molecule of ATP in the presence of a specific enzyme (figure 3.8 (*a*)). The glucose phosphate reacts with GTP to give GDPglucose and pyrophosphate (figure 3.8 (*b*)).

There is a problem at this point. The chemists tell us that this reaction is almost as likely to go one way as the other as there is little in the way of energy change, and reactions only go heavily one way or the other if there is a big energy change. However, the pyrophosphate is readily hydrolysed by a specific enzyme to two molecules of 'ordinary' (ortho) phosphate, with loss of the energy of its linkage as heat:

$$H_4P_2O_7 + H_2O \rightarrow 2H_3PO_4 + \text{much energy}$$

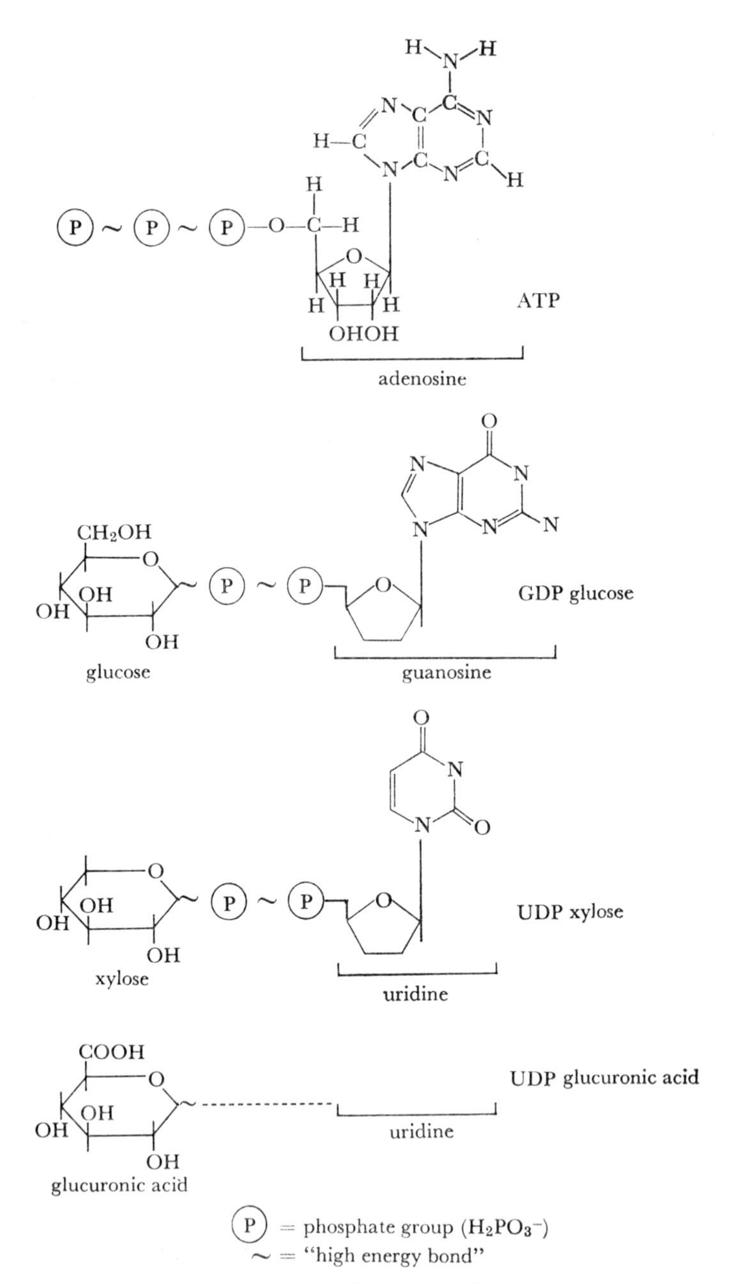

Fig. 3.7 Some sugar nucleotides. The ATP molecule at the top is shown for comparison

(a) + ATP → + ADP

glucose (from respiration) glucose phosphate

(b) GTP

P + P

released

⇌ GDP glucose + PP_i

(c) P ~ P → released

cellulose

(d)

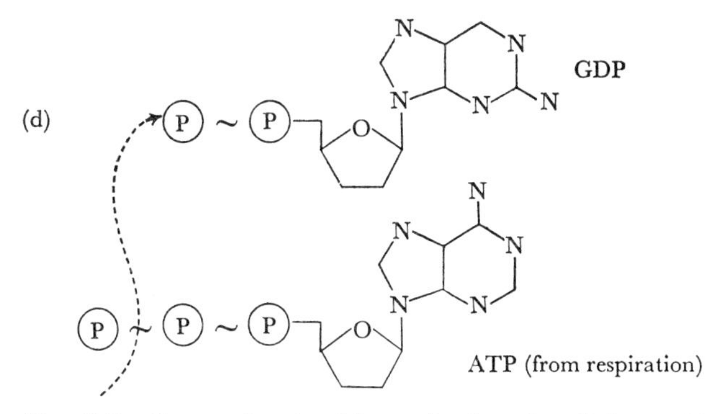

Fig. 3.8. Stages in the biosynthesis of cellulose. Note that ATP is required (steps a and d). The process therefore consumes energy

so reaction (*b*) in fact very readily keeps on producing GDPglucose. Put another way, the PP_i produced by reaction (*b*) is removed by conversion to $2P_i$ so one of the reactants for reversal is no longer available.

This hydrolysis of pyrophosphate is a reaction found almost universally in the biosynthesis of biological polymers (polysaccharides, proteins, nucleic acids) and in the manner just indicated imposes the direction towards synthesis upon them.

The GDPglucose is now able to transfer its glucose to an existing cellulose chain, and GDP is released (figure 3.8(*c*)).

The GDP formed by the transfer of glucose from GDPglucose to cellulose is reconstituted to GTP by phosphate transfer from ATP (figure 3.8(*d*)).

If we co-ordinate these reactions we find that each glucose molecule added to the growing polymer chain involves the loss of at least two high-energy phosphates from the ATP pool: in quantitative terms, therefore, each gram molecule of glucose requires at least two gram molecules of phosphate.

How much respiration in terms of oxygen uptake does this process require? And what proportion of the oxygen uptake shown by growing plant tissues does it represent?

We can take as an example growing seedling tissue of any higher plant, for example beans, excluding that part of the seedling which does not grow, i.e. the food reserves of the cotyledons (or endosperm in the case of an endospermic seed). The growing tissue will increase in both fresh weight, at the expense of water provided externally, and dry weight at the expense of the internal food supply. Let us assume that one gram fresh weight of the growing tissue increases to 11 grams in 100 hours at room temperature, an increase of 10 grams. Let us also assume that the dry weight is one-tenth of the fresh weight; this then increases by 1·0 gram. We will make a final assumption that 90% of the dry weight is cell-wall material, polysaccharide, which therefore increases by 0·9 gram. Although all these figures have been chosen for convenience, the assumptions are reasonable on the basis of known changes in seedlings.

The molecular weight for a hexose sugar (e.g. glucose) is 180, so to make 0·9 gram of polysaccharide from it we are dealing with approximately $\frac{0\cdot9}{180}$ moles $\equiv \frac{5}{1000}$ moles $\equiv$ 5 millimoles of the hexose, and from the argument developed above, some 10 millimoles of phosphate in combination as 'high-energy phosphate groups' is needed to achieve its polymerisation. Since the oxygen molecule contains two atoms, we can say, on the basis of the P/O ratio of 3, that

$$3 \text{ millimoles phosphate} \equiv \tfrac{1}{2} \text{ millimole } O_2$$

and

$$10 \text{ millimoles phosphate} \equiv 1\cdot67 \text{ millimoles } O_2$$

In terms of volume, this is 37 ml oxygen (at NTP 1 millimole $O_2 \equiv 22\cdot4$ ml). Now, we know from experimental observation that a reasonable oxygen consumption rate for growing tissue is 3 microlitres (10^{-6} litres) per milligram dry weight per hour, and it can be calculated, on the assumption that the growth curve of a seedling is approximately logarithmic, that about 120 ml oxygen would be consumed by our seedling tissue in 100 hr. Of this, from the above biochemical considerations, cell-wall synthesis accounts for 37 ml, that is about one-third of the total oxygen.

(*b*) ***Protein synthesis.*** Once again, ATP plays an integral part in the sequence of reactions. Each amino-acid molecule is 'activated' by ATP:

$$\text{amino acid} + \text{ATP} \xrightarrow{\text{enzyme}} \text{amino acid—AMP} + \text{PP}_1$$

Again, direction is imposed by 'squandering' the energy of the pyrophosphate by hydrolysing it to orthophosphate. The activated amino acid is transferred to a molecule of sRNA, a different type for each amino acid (aa):

$$aa_1\text{AMP} + \text{sRNA}_1 \xrightarrow{\text{enzyme}_1} aa_1 - \text{sRNA}_1 + \text{AMP}$$

$$aa_2\text{AMP} + \text{sRNA}_2 \xrightarrow{\text{enzyme}_2} aa_2 - \text{sRNA}_2 + \text{AMP}$$

Respiration reconverts the AMP to ATP:

$$\text{AMP} \xrightarrow{P_i} \text{ADP} \xrightarrow{P_i} \text{ATP}$$

Monomers used for RNA synthesis

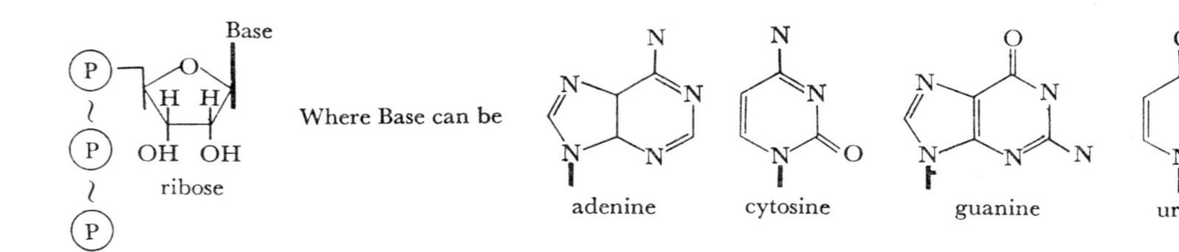

Monomers used for DNA synthesis

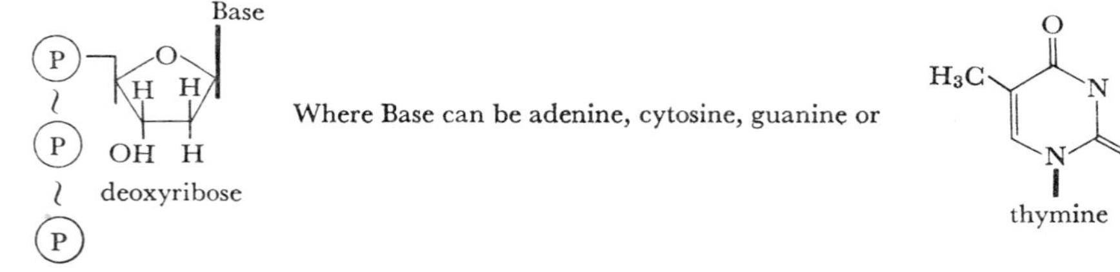

Fig. 3.9. Nucleotides used in the biosynthesis of nucleic acids. Adenine and guanine are called purines, the other bases pyrimidines

Each amino acid–sRNA complex can transfer its amino acid to a growing protein chain:

$$aa_2\text{–}sRNA_2 + aa_3\text{—}aa_4 \xrightarrow{\text{enzyme}} aa_2\text{—}aa_3\text{—}aa_4 + sRNA_2$$

$$aa_1\text{–}sRNA_1 + aa_2\text{—}aa_3\text{—}aa_4 \xrightarrow{\text{enzyme}} aa_1\text{—}aa_2\text{—}aa_3\text{—}aa_4 + sRNA_2$$

The released sRNA molecules are now free to take up more amino acids.

(*c*) ***Nucleic acid biosynthesis.*** These polymers are also built from monomer units. Here the monomers are nucleotides very similar in overall structure to ATP, the high-energy bonds coming from respiration. We have come across another one of these already when considering cellulose biosynthesis, namely GTP, but now it is the base, guanosine, itself which becomes part of the polymer.

For the RNA of higher organisms the bases, and their corresponding triphosphates, are adenine (ATP), cytosine (CTP), guanine (GTP) and uracil (UTP). Their structures are shown

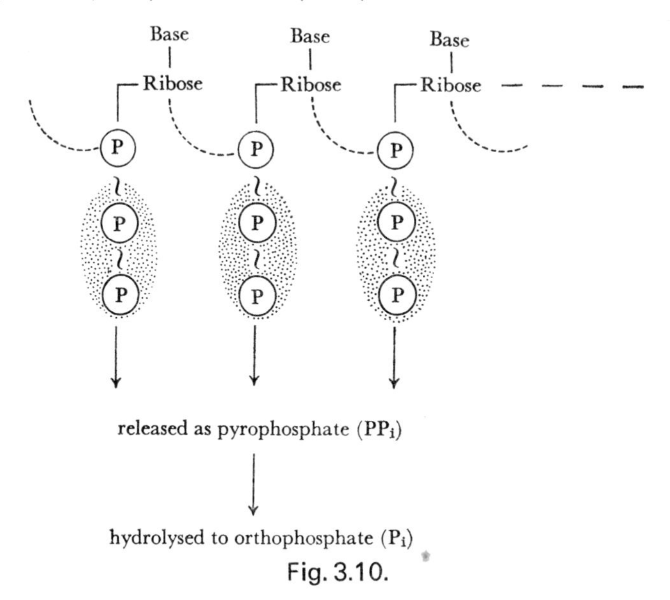

Fig. 3.10.

in figure 3.9. The same bases are used for DNA synthesis, except for uracil which is replaced by the similar compound thymine. The triphosphates used here differ slightly from those used in RNA synthesis in that the sugar ribose which is part of the structure lacks one oxygen atom, thus becoming deoxyribose (figure 3.9).

The nucleic acid chain is made in the way shown in figure 3.10. Once again the release and subsequent hydrolysis of pyrophosphate is an important step.

Summary

In this chapter we have studied some of the constituent processes of growth in considerable depth. We have considered the role of water, of cell-wall synthesis and of protoplasmic changes. Knowledge of the structure of cellular components and of the chemistry of their constituent compounds is obviously necessary, and these have been touched on. Changes in them require energy, showing that growth as a whole is an active process requiring an energy input, derived ultimately from photosynthesis, but locally via respiration from nutrients translocated to the growing centres.

The energy needed for growth does not play a vague or generalised role: it can be pinpointed to particular biochemical events, and we have followed in some detail three of these, namely the biosynthesis of polysaccharide, protein and nucleic acid. For this we have unavoidably needed to go into their biochemistry.

By now you will have realised that a complete understanding of growth can only be achieved by considering its many aspects, e.g. anatomy, physiology, mathematics, chemistry, biochemistry and even genetics and ecology.

Part 2: The internal control

In the previous section we considered growth of organisms, organs and individual cells, and we saw how this occurs in an ordered, co-ordinated fashion with a recognisable pattern. This suggests that some precise control or controls are exerted within the plant at the four levels of organisation – cell, tissue, organ and organism. External factors, such as light, temperature and gravity are, of course, extremely important and we shall see how these profoundly affect the quality and quantity of growth. But these factors intervene in growth by affecting the primary internal controls; we shall discuss this further in Part 3 of this book.

Undoubtedly, the conditions within the plant which affect growth are very complex and almost any change in metabolic behaviour is likely to alter the growth pattern. Clearly, nutrition, including photosynthesis, carbohydrate, fat and nitrogen metabolism, and the mineral salt content, are all ultimately responsible for growth. These are obviously not specific to the growth process though, and exert their influence, in a sense, only indirectly.

Specific control of growth and development is thought to be achieved primarily by special chemicals which influence cell division, enlargement and differentiation. These chemicals are active in minute amounts, often at some distance from the cells which produce them, and we can therefore look upon them as chemical messengers; they are called hormones. Hormones are found in both plants and animals but those in plants (phytohormones) are not synthesised in special organs, which is often the case in animals. Sometimes it is difficult to make a clear distinction between hormones and other growth factors which are active in small amounts, such as the vitamins. For example, roots require thiamin for their metabolism (as co-factors for some enzymes) and hence growth; this substance is provided by the shoot and might therefore fit into the definition of a hormone. But hormones, it should be remembered, have a specific influence on growth and do not act through direct nutritive effects. Thus, thiamin would not properly be considered as a hormone.

There are a number of different types of phytohormone regulating cell growth or division, and which are, furthermore, involved in all the various aspects of development (germination, flowering, dormancy, etc.). In view of recent research it is sometimes difficult to divide them into distinct physiological categories as their effects often overlap. However, the differences and similarities they do possess enable us to group them under the terms auxins, gibberellins, cytokinins and inhibitors. To these groups whose chemical structures are for the most part known, we must add those hormones for which there is good circumstantial evidence but which still remain to be satisfactorily isolated (e.g. flowering hormones). In addition to the hormones, which by definition are produced by the plant itself, we must also consider the growth regulators. These are manufactured chemicals which often resemble the hormones in physiological action and even in molecular structure but they are not naturally occurring. Some of these substances are of great importance in agriculture and horticulture where they have many uses connected with the control of plant growth and development.

Table 1 gives a summary of the effects of the phytohormones which clearly have very diverse actions. All these will be discussed in the following chapters but in Part 2 we will be concerned only with the hormonal control of growth, leaving the development and morphogenetic actions of hormones (e.g. flowering, dormancy, etc.) until Parts 3 and 4.

It is worthwhile now, however, to make some general points. Plants during their evolution have come to use the hormones apparently to perform many tasks. These substances possibly have a single primary 'trigger' action or alternatively they could act differently depending upon the tissue, age, etc. We will discuss this view in Chapter 8 on mechanisms. First, however, let us see how these important chemicals were discovered, what they are, and how they feature in growth control.

4

The Phytohormones

The detection of the plant hormones is an instructive lesson in scientific discovery. The discovery of the three major classes of hormone (auxins, gibberellins and cytokinins) was a logical consequence of investigations into certain physiological processes although these researches did not set out originally to show that such substances are present in plants. By astute observation and careful, stepwise experimentation, however, their occurrence and nature were eventually revealed.

Auxins

Definition of an auxin

The term auxin now includes all those chemicals, from whatever source, which have certain biological activities. Historically, the most important activity is the stimulation of growth of the longitudinal axis of plants 'freed as far as practicable from their own inherent growth-promoting substances'. In fact, this means plants which have had their own source of auxins removed (e.g. the apex). Auxins, then, generally are substances which promote growth of stem or coleoptile sections and decapitated coleoptiles, but in the same concentrations are almost totally without effect on intact plants, at least as far as growth is concerned. They also have a range of other distinctive properties which we shall refer to in Chapter 7.

Discovery

The auxins were the first plant hormones to be discovered. At the end of the 19th century, Charles Darwin and his son studied the bending of grass coleoptiles towards light. They found that the

Table 1

The actions of phytohormones

Process affected	*Cellular basis*	*Auxins*	*Gibberellins*	*Cytokinins*	*Inhibitors (e.g. Abscisic acid)*
Stem growth	Division at apex	No action	Active	Possibly active. Active in lateral buds	Active (dormancy)
	Cell enlargement	Active	Active (but requires auxin)	Sometimes active	Active
Coleoptile growth	Mainly enlargement	Active	Active at some stage	Active at some stage	Active (stops auxin action)
Fruit growth	Division and enlargement	Active	Active	Active	—
Root growth	Division and enlargement	Active	Generally inactive	Generally inactive	—
Leaf growth *	Mainly enlargement	No action	Active	Active	—
Promotion of flowering *	Differentiation	No action	Active (e.g. in L.D.P.)	Generally inactive	Sometimes active
Sex expression	Differentiation	Favours femaleness	Favours maleness	No action	—
Root initiation	Division, enlargement, differentiation	Active	Inhibits	No action	—
Bud formation (e.g. on calluses, or cuttings)	Division, enlargement, differentiation	Sometimes active	Sometimes active	Active	—
Breaking of dormancy in seeds and buds*	Division and enlargement	No action	Active	Sometimes active	—
Lateral bud inhibition	Division and enlargement	Active	Enhances auxin action	Antagonises auxin	—
Cambial division	Division and differentiation	Active (with gibbs)	Active (with auxins)	?	—
Protein synthesis	—	Active	Active	Active	Sometimes active
Leaf abscission	Differentiation	Active	No action	No action	Sometimes active
Delay of senescence	—	Active	Active	Active	—

This table summarises the effects of hormones and, necessarily, includes some generalisations. Different species do not respond equally nor do the various growth substances invariably act as shown.

tip of the coleoptile perceived the light and then exerted some 'influence' which controlled the differential cellular expansion in the growing region several millimetres below the apex. In the 1920s and 30s this work was extended, particularly by Boysen-Jensen and Went, who set the concept of a plant hormone firmly on its feet. They noted that elongation of cells in the expanding zone of oat (*Avena*) coleoptiles almost stopped when the tip was removed, but was restored when the latter was replaced by means of gelatin. Growth occurred only on the side immediately beneath the tip, if this was replaced eccentrically, and the coleoptile bent as though it had been illuminated from one side (figure 4.1).

Isolation of a known chemical

These results provided firm evidence that the coleoptile apex produced some mobile substance(s) which controlled growth of other cells of the coleoptile, i.e. a hormone. Plant physiologists then set about the task of collecting and identifying it. The hormone was found to diffuse from excised coleoptile tips into small blocks of agar jelly which, when placed on a decapitated coleoptile, restored its growth to a level comparable with that in the intact organ. A growth curvature resulted when the agar was applied eccentrically. This was used to measure relative concentrations of the growth substance, for the angle of curvature depended on the concentration in the agar. However, attempts to isolate the active substance from coleoptile tips were unsuccessful as the amounts present are very small.

Other sources of the chemical which could control coleoptile growth were therefore sought. Eventually, active substances were obtained from human urine, maize oil, malt, yeast and from the mould *Rhizopus*. These substances were named auxins, the word derived from the Greek, *auxein*, to increase. After thorough chemical investigation it was established that the auxin activity from those various sources was due to one chemical, identified as β-indolylacetic acid, now usually abbreviated to IAA (figure 4.2).

The auxins of higher plants

IAA has been found in a number of species (e.g. bean stems, potato tubers, sugar-cane, pineapple) but its presence in many

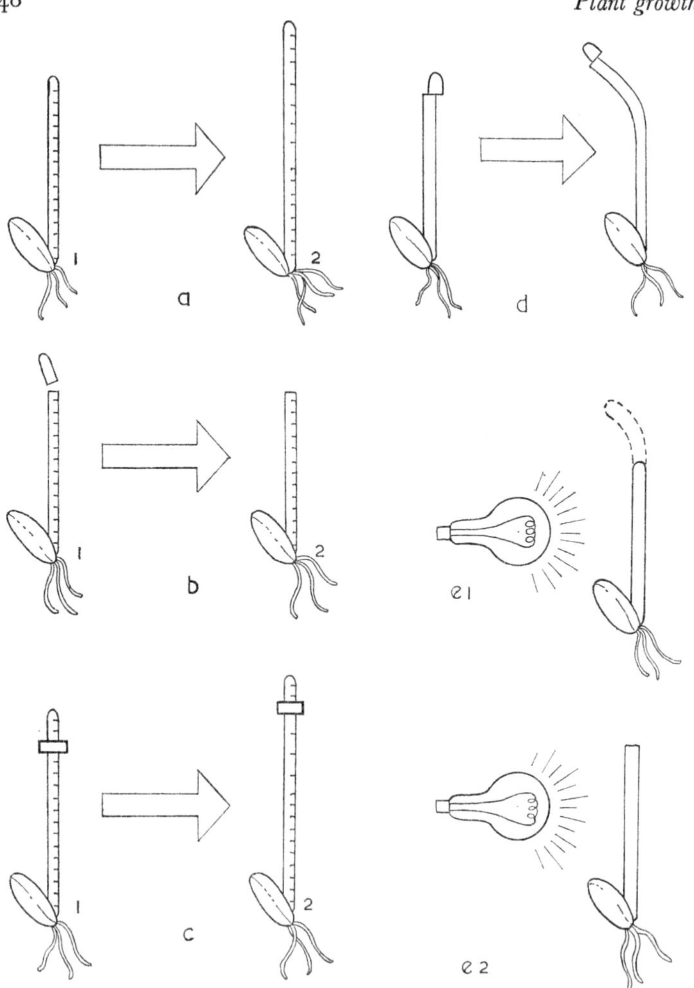

Fig. 4.1. The growth of coleoptiles. Note that the region of growth is several millimetres below the tip where the markings become most spaced out (*a*, 1 and 2). The growth is stopped when the tip is removed but continues if it is replaced (*b* and *c*). An eccentrically-replaced tip causes growth only on one side (*d*), and bends as though illuminated (*e*, 1). No bending towards light occurs when the tip is removed (*e*, 2). Thus, the coleoptile tip controls growth of the cells below

growing tissues has not yet been established. Thus, although for many years IAA was considered as the sole naturally-occurring auxin, its exclusive position has often been challenged.

We now know of other derivatives of indole which occur in plants and which qualify as auxins (figure 4.2). Indole acetonitrile (IAN) was first extracted in 1952 from cabbage leaves and has also been found in other species belonging to the *Cruciferae*. This substance causes excised oat coleoptiles to elongate but does not produce curvatures when applied to the top of decapitated coleoptiles. This is probably because of the poor transportability of IAN within the tissues (see below). There is good evidence that IAN is converted in many plants to IAA. The plants which can perform this conversion (e.g. oat) are sensitive to applied IAN, but those which cannot (e.g. pea) are insensitive. Other indoles with auxin activity which have been found in plants are also shown in figure 4.2. Some of these may not be auxins in their own right, but like IAN, must first be changed to IAA.

In plants, the indole auxins are thought to be synthesised from the amino acid, tryptophane, through various intermediates (figure 4.2). IAA and IAN can also be derived in some plants (especially in *Cruciferae*) from hydrolysis of complex glycosides called ascorbigen and glucobrassicin.

Distribution in higher plants

A comprehensive survey of the auxins in etiolated oat seedlings showed the greatest quantities in the root and shoot apices and the lowest in the non-growing regions of the plant. In the broad-bean, too, relatively high concentrations were found in the apices and young leaves (figure 4.3). In general, then, young, actively growing tissues, such as root and shoot apical meristems, and enlarging leaves seem to be the major centres of auxin synthesis. In the leaves, this continues even for a period after growth has stopped. We should note here that the concentrations involved are 'tiny', perhaps of the order of 10 μg (i.e. 1×10^{-5}g) per kilogram of plant fresh weight.

Although we have spoken above only of 'auxins', three different states of these substances, in fact, exist in plants. These are *extractable*, *diffusible* and *bound* auxin. A reliable picture of the

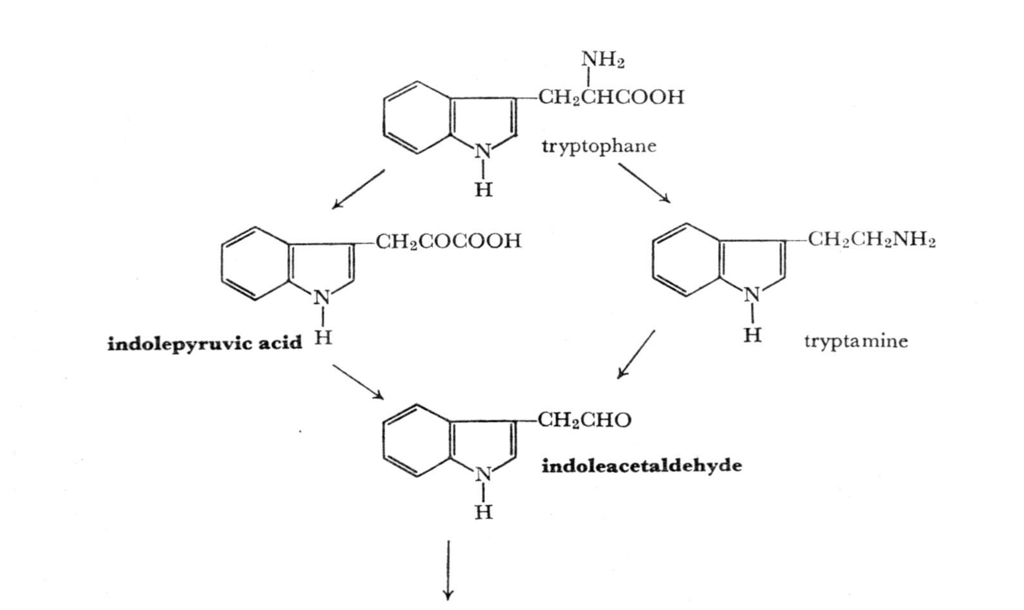
NH2
CH2CHCOOH
tryptophane
N
H
CH2COCOOH
N
H
indolepyruvic acid
CH2CH2NH2
N
H
tryptamine
CH2CHO
N
H
indoleacetaldehyde

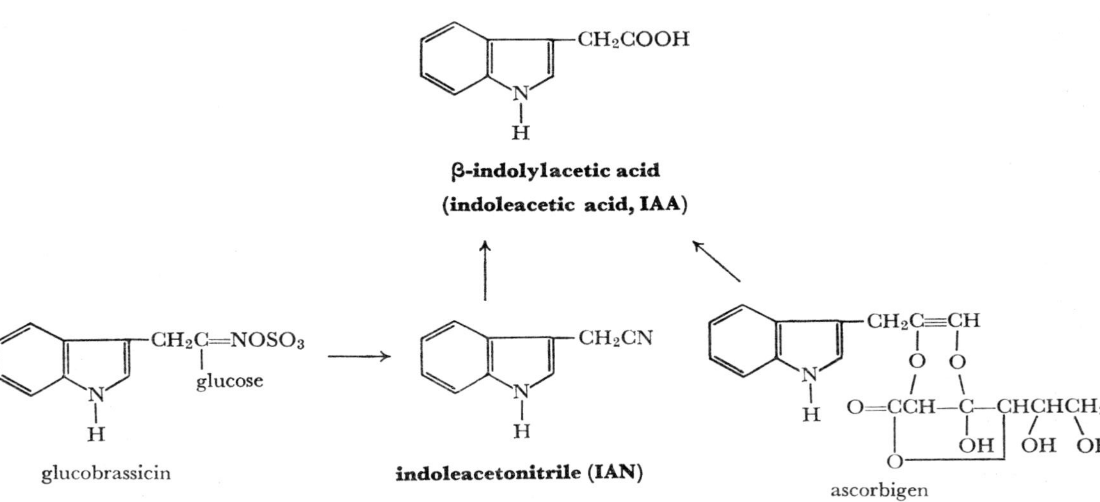

Fig. 4.2. The probable biosynthesis of indole auxins. The chemicals named in bold type are those which are considered to be the naturally-occurring auxins

auxin content can often be obtained by measuring the amount of auxin diffusing out of a cut tissue into, for example, agar jelly, or by extracting with organic solvents. Unfortunately, though, the former method is sometimes unreliable because enzymic oxidation of auxin can occur at exposed tissue surfaces. Perhaps an even

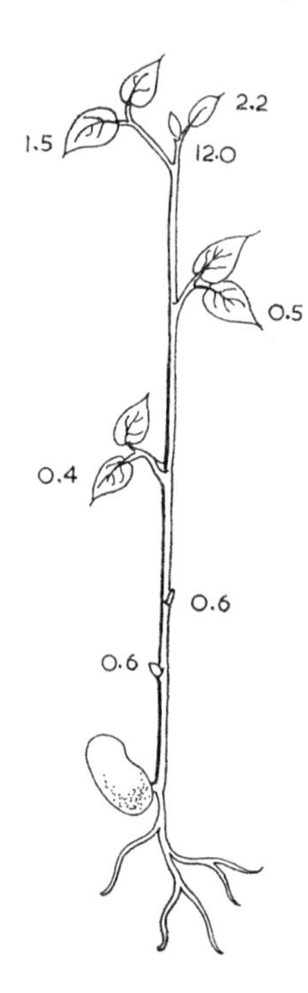

Fig. 4.3. The distribution of auxin in leaves and buds of a young broad bean plant. The numbers are units of diffusible auxin per hour. (after Thimann and Skoog)

greater danger is that a completely misleading picture of the real auxin content might be obtained. This is because the auxin in the tissue might be quite firmly bound on to cellular components such as protein, and is therefore non-diffusible and non-extractable. The leaves of the pineapple plant, for example, are poor sources of free auxin but are rich in auxins which are slowly liberated from a tissue homogenate or are more quickly released after treatment with proteolytic enzymes (e.g. ficin, trypsin). The free auxin in the apex of the pineapple plant is even possibly produced by dissociation of the auxin–protein complex in the leaves.

Seeds are rich in both free and bound auxins and in auxin precursors, which are substances giving rise to auxins. The endosperm of oat grains, for example, furnishes a precursor which is translocated to the coleoptile apex. Auxins in seeds are particularly important in the control of fruit growth, as we shall see later.

Movement of auxin in plants

To qualify as hormones the auxins must exhibit some degree of movement within the plant, i.e. they must act at distance from their source. In the coleoptile, mobility is obvious for we have seen how cell enlargement in the growing region is controlled by auxin from the apex. Here, auxin is moving in a basipetal direction, i.e. towards the morphological base. This is, in fact, an important characteristic feature of auxin, or at least of IAA. With some qualifications, we may say that it moves only in this direction and therefore exhibits a distinct 'polarity'.

The plant physiologist Went first experimented on this phenomenon in 1928, followed later by other workers. Movement of auxin from the tip into the body of the coleoptile was found to be prevented if a short section of coleoptile was cut, inverted and interposed between the tip and the basal part but it could continue provided the excised section was placed the 'correct' way around (figure 4.4). In later experiments excised coleoptile sections were arranged between two blocks of agar, one, the donor block, containing auxin and the other, the receiver block, with none. Auxin moved only from the block at the apical end of the section to the basal end no matter how the blocks and sections were arranged with respect to gravity (figure 4.4). Moreover, this basipetal movement continued against a concentration gradient. Such experiments were used to calculate that auxin has a transport rate through living cells of about 10 mm per hour. This rate, and the movement against a gradient rule out simple physical diffusion. Living, actively metabolising cells control the movement. This is also demonstrated by the inhibitory effect upon the movement of low oxygen concentrations and of respiratory inhibitors.

Polar movement takes place in the parenchyma of the coleoptile and does not require the presence of the vascular system, although auxin does move down the phloem of stems. Polarity is well defined in young tissues but it may disappear as cells mature. For example, it is almost absent in the lowermost regions of stems

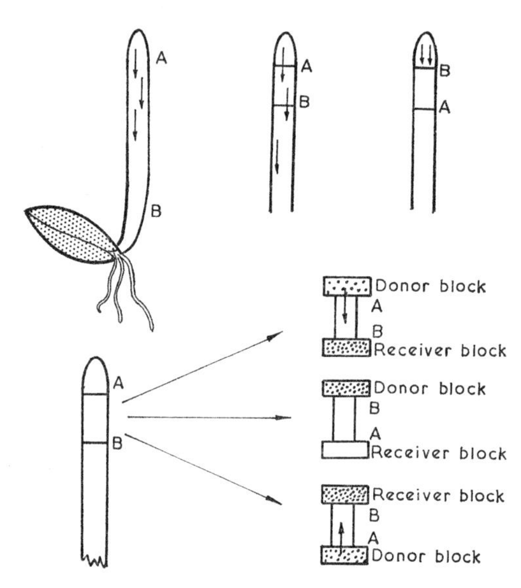

Fig. 4.4. Polarity of auxin movement. Auxin moves from the apical end (*A*) to the basal end (*B*). This movement stops when a coleoptile section is removed and inserted with the basal end (*B*) uppermost. Similarly, movement of auxin from agar blocks occurs only from apical end to basal end of coleoptile section. (The arrows indicate auxin movement)

and in the older petioles. In these cases, auxin transport can take place with the stream of food materials in the plant, even from the mature leaves to young leaves near the apex – the reverse of its more usual direction. Moreover, many auxins can be taken up by the roots, from an external supply, and move in the transpiration stream of the xylem.

The polarity of auxin movement (and indeed of other sublate 3 *Electron icrographs showing cell ganelles of Jerusalem tichoke.*

:) *Young cell:* The ıcleus (nuc) :cupies a large part `the cell and ɒntains a nucleolus ı) and chromosome aterial (c). There ·e numerous small ıcuoles (v) which ill fuse as the cell ows to give the huge ll vacuoles shown in). The cytoplasm ɒntains mitochondria n), golgi bodies (g), c; the tiny black anules are bosomes. asmadesmata (p) averse the walls agnification ×5,500.

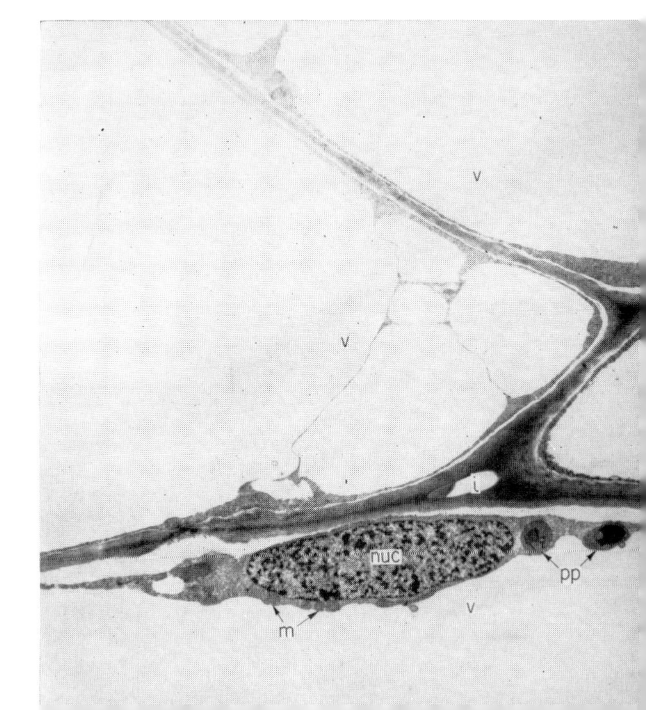

) *The junction of four ture cells;* Note the inness of the ripheral cytoplasmic ·er in relation to the e of the vacuoles, ıich are also ıversed by toplasmic strands. oplastids (pp), tochondria, a cleus, and an inter- lular space (i) can seen. Magnification 1,240.

(a)

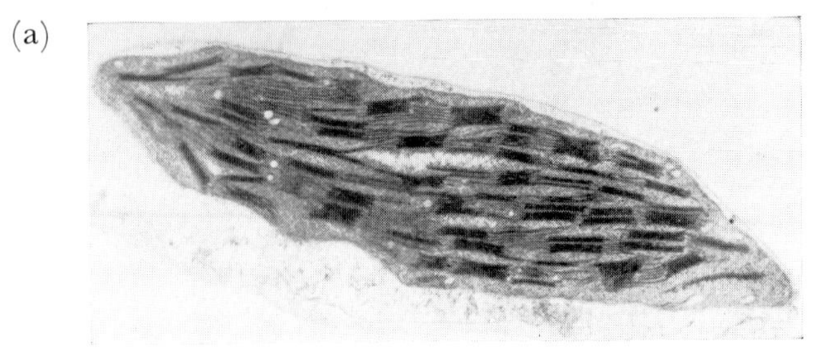

(b)

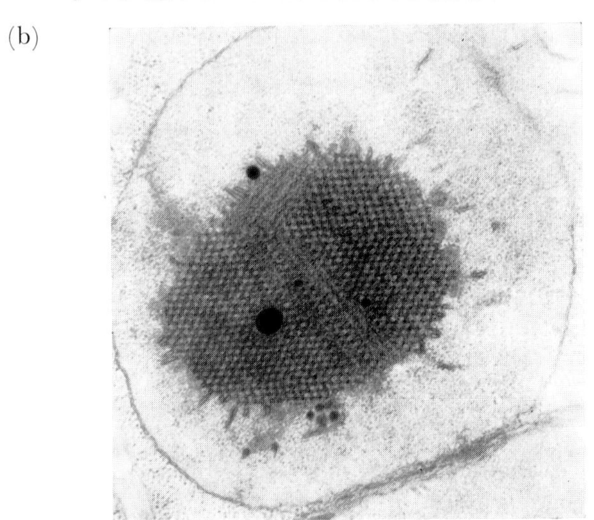

(c)

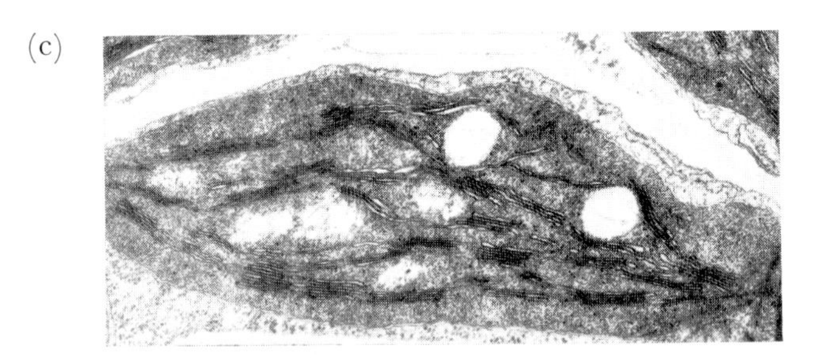

Plate 4 *Effect of light upon differentiation of bean chloroplasts.*
(*a*) Section of a normal chloroplast from a 15-day old plant. Magnification ×23,600.

(*b*) When a plant is grown for a similar period in total darkness, it does not go green, and aberrant chloroplasts, called etioplasts, are formed. Magnification ×27,500.

(*c*) After exposure of the etiolated plant to light for 72 hours the etioplasts have differentiated almost completely into normal chloroplasts. Magnification ×42,000. The beginning of this process can be detected within a few minutes of exposure.

Plate 5 *The use of synthetic auxin* (2, 4-D) *as a selective herbicide.* This field of oats was sprayed with the auxin except for a test strip; here the weed (charlock) has flourished. The spraying has killed the charlock but left the oats unharmed.

Untreated　　Treated with Phosphon

Plate 6 *The effect of a growth retardant (Phosphon) on growth of chrysanthemum.*
Note the great reduction in stem length with little effect on the general health and flowering of the plant.

stances) cannot yet be explained. Various possibilities have been suggested, such as protoplasmic streaming and bioelectric potentials, but none can be considered proved. Nevertheless, there is no doubt that polarity plays a very important role in the developmental effects of auxins (see, for example, Chapter 7).

Auxin regulation in the plant

We shall see later that the concentration of auxin is extremely important in growth control. Clearly then, in order to function properly as a growth-controlling substance within the plant the amount of auxin in the tissues must be regulated. Evidently, plant tissues can do this, for although the apex or the young leaves may continue to synthesise auxin over a fairly long period the concentration in the growing regions remains constant or even falls.

The internal level of active auxin can be affected in three ways. As we have already mentioned, auxin may be bound on to protein and thus inactivated. Secondly, certain inactive derivatives are formed by reaction with other chemicals. For example, when IAA enters pea roots from the outside medium it does not remain as free auxin but is transformed into a peptide, indole aceoaspartate, by combination with aspartic acid (figure 4.5). This is perhaps a detoxification system for removing excess auxin, and it may be of widespread significance in plants. Other IAA conjugates, for example, with sugars to form glycosides, have also been found. Beside being inactivated in these ways auxin can also be oxidised by enzymes called IAA oxidases. These enzymes have been found in many tissues and are thought to be important in the regulation of auxin levels. Further, light can apparently activate these enzymes and thus possibly influence auxin concentrations.

Gibberellins

Definition

The gibberellins are hormones which strongly promote plant growth, especially of stems. Unlike the auxins, though, they increase growth when applied to intact plants, sometimes by a factor of four or five times, but can be almost inactive on stem or coleoptile sections. Thus, although both these hormones affect growth, they seem to act in a rather different manner. The

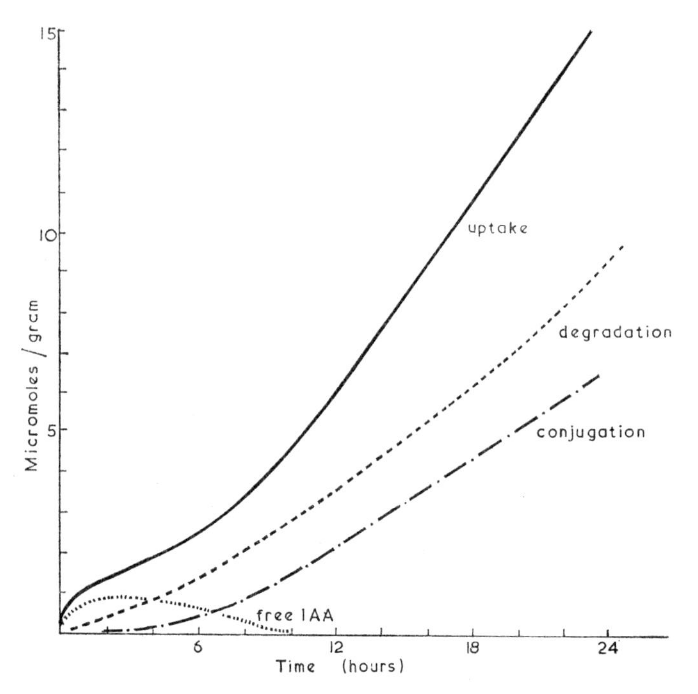

Fig. 4.5. The fate of IAA in pea roots. Although uptake of IAA rises steeply over 24 hours, free IAA is detectable within the roots for only a few hours. After about 6 hours some IAA becomes conjugated to form the indoleacetoaspartate and some of it is degraded

gibberellins, furthermore, have striking effects on flowering and dormancy – aspects of development which are more or less unaffected by auxins. Further differences between these two classes of phytohormone will become apparent when we discuss growth control, and development.

Discovery

When European and American plant physiologists were first investigating the auxins (1920–35) Japanese biologists were engaged in studies which led to the discovery of the gibberellins.

The 'bakanae' (foolish seedling) disease of rice is a fungus disease which was first recorded in the 19th century and is characterised by the greatly elongated stems of the rice plants, which give poorer yields. Plant pathologists in Japan showed that the pathogen was the fungus *Gibberella fujikuroi* (= *Fusarium monoliforme*). They also discovered that the same symptoms of overgrowth could be produced by applying to uninfected plants some of the cell-free filtrate of the medium in which the pathogen had been cultured. The disease symptoms, then, were caused by something which could diffuse from the mycelium even into a culture medium. Eventually, the active material was purified and crystallised, and was called gibberellin A after the genus of fungus.

Partly because of the Second World War, these findings received little or no attention in other countries until the early 1950's when scientists in Britain and the U.S.A. became interested. In the succeeding years gibberellin A (GA) was further purified into three constituents, GA_1, GA_2 and GA_3. GA_3 was shown to be identical to gibberellic acid which had, in the meantime, been isolated from fungal cultures. Higher plant tissues were then examined and were also found to contain gibberellins.

The chemistry of the gibberellins

From the formulae shown in figure 4.6, the gibberellins are obviously very different from the auxins. The variation among the gibberellins (i.e. degree of reduction, hydroxylation) are matched to some extent with their physiological differences. For example, some of the gibberellins are effective only in certain species or in certain developmental processes. Even where a number of different gibberellins are effective in the same tissue, there are invariably differences in the amount of activity.

We might expect that the biosynthesis of such complex molecules is a complex process, and indeed it is. The synthetic pathway is linked with those for the production of steroids and

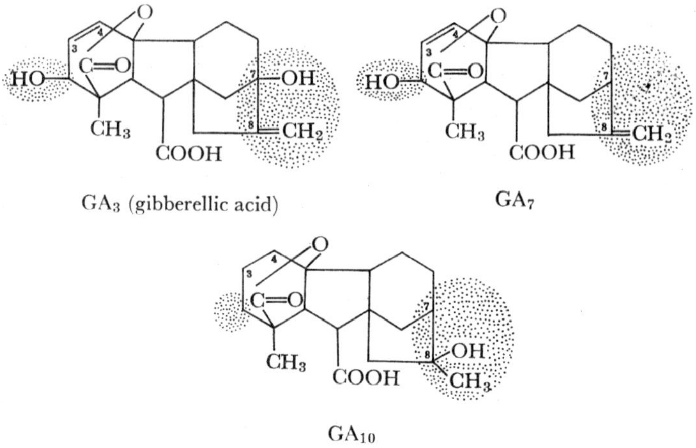

GA_3 (gibberellic acid) GA_7

GA_{10}

Fig. 4.6. Some gibberellins. The parts of the basic structure where the differences occur are shaded

terpenoids. Figure 4.7 shows in very brief outline how the key intermediate mevalonic acid is the starting-point for the production of these chemicals.

Occurrence, transport and regulation in higher plants

At the time of writing twenty-three different gibberellins have been described, some much more common in plants than others. They are all written in abbreviated form as GA_1, GA_2 . . . GA_{12}, etc. Many gibberellins are confined either to the fungus or to flowering plants, but others, including GA_3, are found in both. Gymnosperms, ferns, algae and bacteria all yield gibberellin-like substances but most of these have not been chemically identified.

Some of the first sources in higher plants were fruits and seeds, especially young developing seeds (e.g. French bean and *Echinocystis*, from the cucumber family). In immature bean seeds the concentration of gibberellins is about 0·025 μg per gram fresh weight. Seeds also contain 'bound' gibberellin (i.e. a gibberellin-protein complex) and the free hormone can be liberated by acid or enzymic hydrolysis. When radioactive gibberellin, synthesised chemically, is fed to maturing seeds the binding can be followed

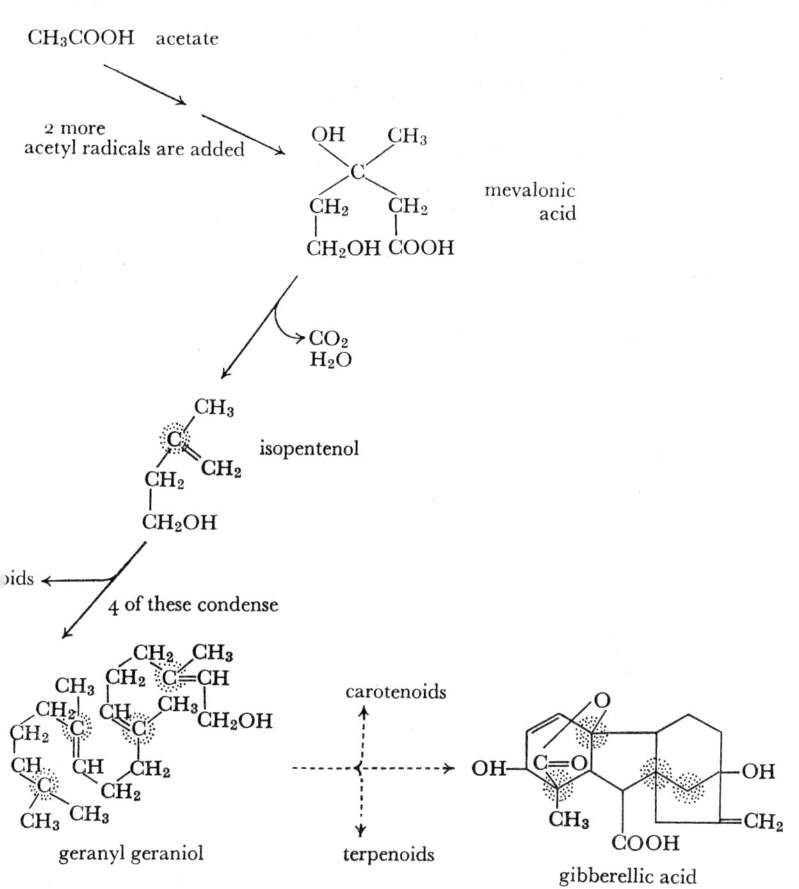

Fig. 4.7. An outline of the biosynthesis of gibberellins

directly and the labelled gibberellin is found to be released once more when the seeds later germinate. Gibberellins also form conjugates with sugars and amino acid (i.e. to give a glycoside and a peptide). In adult plants gibberellins are formed by young leaves at the apex and also in the roots from which they move to the shoot in the xylem sap (e.g. in sunflower and pea). Active gibberellin synthesis (as opposed to release from a bound form) also occurs during germination of certain seeds, for example, barley and wild oats.

Gibberellin transport within the plant has not been investigated as intensively as has the movement of auxin, but they do not appear to exhibit any well-defined polarity. They move into apices from the cotyledons (e.g. in bean seedlings) and they also move upwards into the xylem sap of some plants, as we have already mentioned. Gibberellins can actually be collected from this sap which exudes from the stump that remains when the shoot is cut off.

Little is known about the eventual chemical fate of gibberellins in the plant. They can certainly be broken down but the products of their metabolism are not really known. The formation of gibberellin conjugates which has been described above may be involved in the 'detoxification' mechanism for removing excess hormone.

Cytokinins

Definition

The cytokinins are substances which act primarily on cell division and have little or no effect on extension growth. Thus, they differ markedly from both auxins and gibberellins. For now, it will suffice to keep just these differences in mind, but more will become clear in subsequent chapters.

Discovery

In 1954–56 in the U.S.A. Skoog and his collaborators were working on tissue cultures obtained from pith cut from tobacco stems. They found that the pith cells exhibited cell division for only a short time when grown on agar containing sugar, mineral salts, vitamins, auxin, etc. However, division continued when yeast extracts or coconut milk were added to the medium and eventually a large mass of parenchymatous cells called a callus was formed (see Chapter 10). The biological activity of the extracts was found to be due to a purine (see page 40) but the more common purines such as adenine and guanine were ineffec-

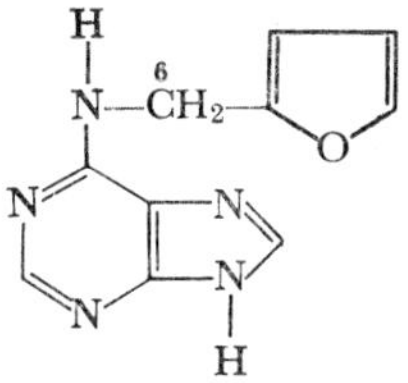

Kinetin (6-furfuryl adenine)

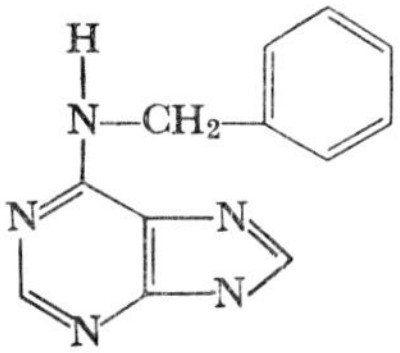

Benzyl adenine

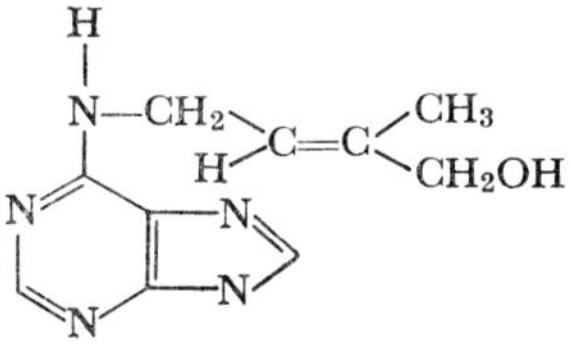

Zeatin

Fig. 4.8. Cytokinins

tive when tested. When the nucleic acids DNA and RNA were tried as possible alternatives to yeast extract, a sample of 'pure' DNA (extracted from herring sperm) proved to be active. Subsequently, it was discovered that the active chemical was really a breakdown product of the DNA obtained after storage at room temperature for some months or by heating in an autoclave. This was isolated, purified and identified as 6-furfuryl aminopurine (or 6-furfuryl adenine) (figure 4.8). Because of its physiological activity in inducing cytokinesis (i.e. actual *cell* division) it was given the non-chemical name, kinetin. Other adenine derivatives, e.g. benzyladenine have activity similar to kinetin, as do various urea derivatives, many of which do not occur naturally. All these chemicals which have the same biological action as kinetin are now called cytokinins.

Cytokinins in higher plants

Now, excised pith shows cell division for long periods when pieces of vascular tissue are left attached or when phloem diffusates or coconut milk is added. Presumably, then, these additives from a higher plant source also contain cytokinins. This has been con-

caffeic acid

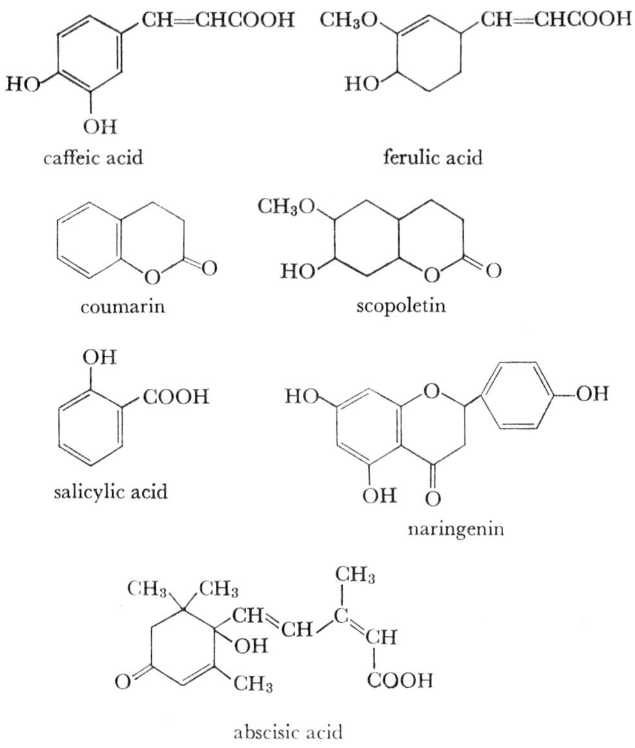

ferulic acid

coumarin

scopoletin

salicylic acid

naringenin

abscisic acid

Fig. 4.9. Some inhibitors

firmed and other higher plant tissues have been found to contain these substances. They have been isolated from young, developing fruits (e.g. apple, peach, tomato) and germinating seeds (lettuce, barley). They are apparently synthesised in the root apex from which they move, in the xylem, up into the stem, and are contained in the 'bleeding sap' which exudes from cut stems of grapevine, sunflower and tobacco.

The structures of only a few naturally-occurring cytokinins are known. Zeatin (the cytokinin from maize seeds) for example, is a complex derivative of adenine, and is, therefore, similar to kinetin itself.

As just mentioned, cytokinins can certainly move in the xylem from the roots to stems and probably also to the leaves and fruits. However, when applied to intact organs such as leaf or stem surfaces they (or at least kinetin) are apparently immobile; they can be transported when applied to decapitated stems.

Inhibitors

Numerous chemicals having inhibitory actions have been extracted from plants and many are particularly important in developmental processes such as dormancy, as well as in general growth.

The formulae of some of these substances are shown in figure 4.9. The inhibitors clearly fall into diverse chemical groups. Phenolic acids, lactones and flavonoids (e.g. caffeic acid, coumarin and naringenin respectively) are possibly the best represented chemical categories. Many of these substances inhibit coleoptile and stem growth and seed germination.

An important recently-discovered inhibitor is abscisic acid (abscisin II and dormin have been used as alternative names). This is a sesquiterpenoid and is therefore related to the gibberellins which are diterpenoids. Abscisic acid can be synthesised in leaves and then translocated in the phloem to stem apices (Chapter 12). It is also found in some seeds and stems.

Many other inhibitors have been allocated physiological roles even though their chemistry is largely unknown. The 'inhibitor β', for example, is of fairly widespread occurrence; it is a mixture of substances including phenolic and abscisic acids and possibly others.

Summary

The four classes of phytohormone – auxins, gibberellins, cytokinins and inhibitors – are chemically very different from each other. They are synthesised in various parts of the plant from which they move either in a basipetal, polar fashion (e.g. the auxins) or in the vascular system (e.g. cytokinins, gibberellins and the inhibitor abscisic acid).

Various devices exist for the regulation of hormone levels in the plant such as enzymic oxidation (e.g. of auxin) or the formation of chemical conjugates (e.g. auxins and gibberellins).

It would be wrong to suggest that these are the only phytohormones. There is evidence that others exists (e.g. flowering hormones) and some have been postulated in order to explain certain developmental processes (e.g. shoot formation).

5

Other growth regulators

In addition to the native growth-controlling substances there are many synthetic chemicals which exert powerful effects on plant growth. Some of these are, of course, nutrients (e.g. nitrates, potassium and phosphorus compounds) and others, such as various weedkillers, are poisons. We are not concerned in this chapter with chemicals of these sorts but rather with those which, in relatively very low concentrations, act specifically on the growth processes. Such substances often act in a similar manner to the endogenous hormones or are, on the other hand, specifically antagonistic to them.

Synthetic auxins

Not long after IAA had been identified as a natural auxin, other substituted indoles, indolepropionic acid and indolebutyric acid, were found to have similar activity. Subsequently, 1-naphthalene acetic acid (NAA), 2-naphthoxyacetic acid, phenylacetic acid and various chlorinated phenoxyacetic and benzoic acids were found to be active as auxins. Certain dithiocarbamates, which, we should note, are not aromatic, also have auxin-like properties (figure 5.1).

All the effects of the native auxins on growth and development can be found among the synthetic ones although any single synthetic auxin may not possess such a wide range of activity as IAA. Since many do not show polar transport they may be inactive in, say, the coleoptile curvature test or in the various correlative actions of auxin (Chapter 7).

The synthetic auxins also differ from the native auxins in being somewhat resistant to enzymic attack within the plant, and in a

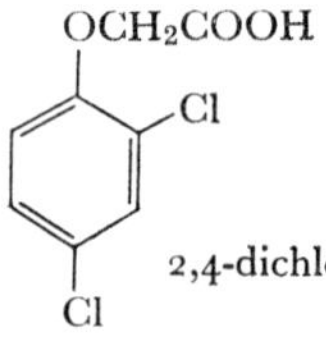

OCH_2COOH, Cl, Cl

2,4-dichlorophenoxyacetic acid
(2,4-D)

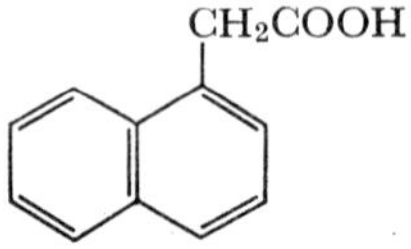

CH_2COOH

1-naphthalene acetic acid
(NAA)

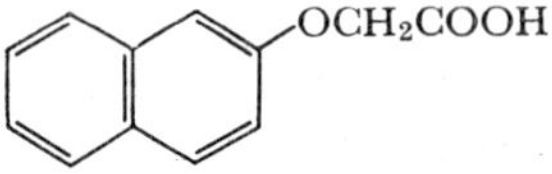

OCH_2COOH

2-naphthoxyacetic acid

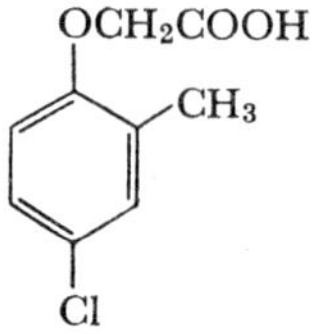

OCH_2COOH, CH_3, Cl

4-chloro-2-methylphenoxy acetic acid
(MCPA)

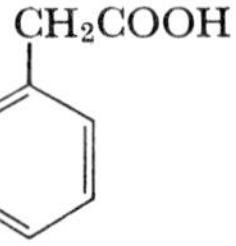

CH_2COOH

phenylacetic acid

COOH, Cl, Cl, Cl

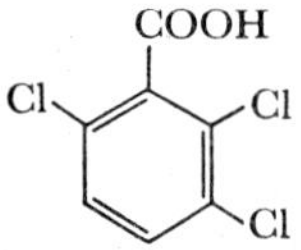

2,3,6-trichlorobenzoic acid
(2,3,6-TBA)

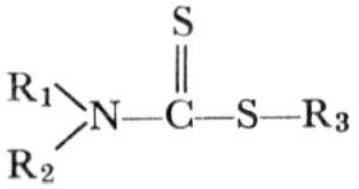

$R_1R_2N{-}C(=S){-}S{-}R_3$

R_1 = eg. CH_3
R_2 = eg. H
R_3 = eg. Na

dithiocarbamates
(the carbamates have O instead of S)

Fig. 5.1. Synthetic chemicals with auxin activity. Some are important agricultural compounds

few cases they may persist for up to several weeks after application. This relative persistence is one reason why the synthetic auxins are of such value in agriculture and horticulture. Sometimes, applied chemicals are degraded by the plant but to a product which still has auxin activity. An interesting example of this is the chemicals of the chlorophenoxyalkyl carboxylic acid series (figure 5.2). Compounds having an even number of carbon atoms between the carboxyl group and the ring have little or no auxin action, whereas those with an odd number are active. This is because in the former case the side chain can be completely oxidised until only the corresponding inactive phenol remains. This oxidative process occurs by β-oxidation, in which carbons are removed two at a time. A carbon chain with an odd number of carbon atoms between the ring and the carboxyl cannot be degraded by this process further than the acetic acid. The resulting chemical – chlorophenoxyacetic acid – is still active as an auxin.

The use of synthetic auxins in agriculture and horticulture is widespread, and they are second in importance as agricultural chemicals only to the fertilisers. Their most common application is in weed control as selective herbicides. The selective, toxic action of these chemicals was first noted in the case of naphthalene acetic acid and later, in about 1942, the more effective methylchlorophenoxyacetic acid (MCPA) and 2, 4-dichlorophenoxyacetic acid (2, 4-D) were discovered. In these preliminary trials it was found that broad-leaved dicotyledons, into which category most of the weeds fell, succumbed to spray treatments, exhibiting distorted growth followed by a quick death, whereas the monocotyledonous cereal crops (wheat, oats, etc.) were unaffected. Over 100 million acres each year are now treated with auxin herbicides, mostly 2, 4-D and MCPA, to control weeds among cereals, sugar-cane and similar plants (Plate 5). The lawn weedkiller preparations are also based on these auxins.

The benefits deriving from the use of these herbicides in agriculture are substantial. Weed contamination can seriously reduce yields and decrease the value of a crop. In some parts of the world the use of auxin herbicides has increased wheat yields

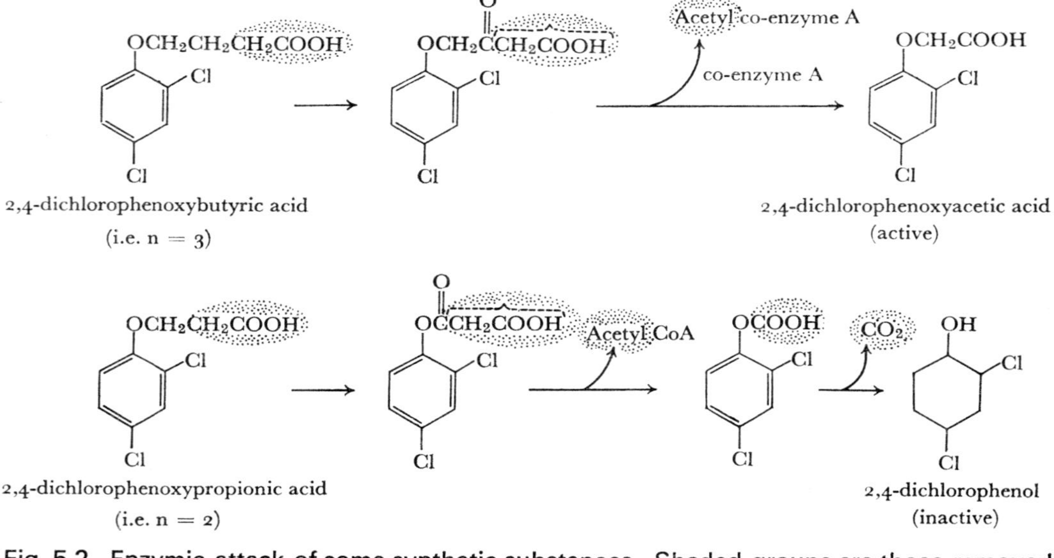

Fig. 5.2. Enzymic attack of some synthetic substances. Shaded groups are those removed by enzymes

by as much as one-third and in California a 75% increase in the rice crop has been obtained!

The mechanism of selective action is not fully understood. Of course, any auxin in relatively high concentration, including IAA itself, will disturb the normal growth pattern, but the latter rarely does this permanently as plants, having a means of controlling hormone levels (see page 55), can destroy the excess. The unnatural auxins 2, 4-D and MCPA, on the other hand, cannot be so easily dealt with.

The disturbance of normal cellular growth and activity leading eventually to death is not so readily brought about by equivalent concentrations in monocots as in dicots. This may reflect real metabolic differences between the two groups but an important contributory factor is also the penetration of the auxins into the leaves. We might expect less of the chemical to fall and remain on long, narrow leaves, held at an acute angle to the stem, than on the broader, horizontal leaves. There is also some evidence that in the less sensitive plants the auxins are not so rapidly translocated from the leaves to the other organs.

The synthetic auxins have many other important practical uses. In orchards, NAA, 2, 4-D, or 2, 4, 5-T are employed to increase fruit set and fruit size or to prevent pre-harvest drop (e.g. in apples and pears). Paradoxically, the same auxins might be used to thin out surplus fruit by inducing fruit drop so that fewer but sturdier ones remain. This is rather similar to the use of auxins as defoliants in agriculture or, more recently, in warfare to expose enemy troops in jungles.

Synthetic auxins are used to control flower formation in pineapple (page 158), thus achieving synchronisation of fruiting. Rooting, which is important in plant propagation, is also promoted by certain synthetic auxins, especially in those plants which would otherwise only poorly form roots on stem cuttings.

These many uses of synthetic auxins are nothing more than practical applications of the roles that native auxins can play. It is important to remind ourselves that our knowledge of auxin action derives ultimately from the apparently 'academic' scientific curiosity of the Darwins as to why grass seedlings bend towards the light!

The anti-auxins

These are chemicals which impede the action of auxins. Thus, auxin-induced growth and leaf abscission may be prevented, or processes which auxins inhibit are promoted (e.g. root growth).

One substance often considered as an anti-auxin is maleic hydrazide (figure 5.3). This has had practical use to induce branching or to reduce growth (e.g. of lawns). However, the substance prevents bud growth in potatoes which paradoxically, auxin itself also does.

Many of the other anti-auxins are chemically similar to the auxins but lack some essential structural feature of the latter. For example, 2, 6-dichlorophenoxyacetic acid and 2, 4–dichloroanisole (figure 5.3) are respectively deficient in the correct ring structure and side chain (see page 101). However, some chemicals which on similar grounds should be anti-auxins are found not to act so.

Many plant physiologists have nevertheless considered that anti-auxins compete with the auxins for some common site within the cell thus blocking it for auxin action. This view is supported by the fact that potential effects of the anti-auxins can

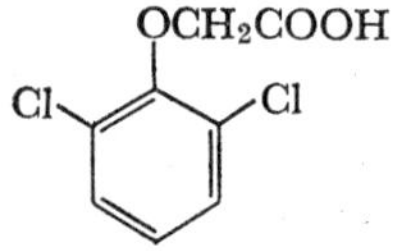

OCH_2COOH, Cl, Cl

2,6-dichlorophenoxyacetic acid

$OC(CH_3)_2COOH$, Cl

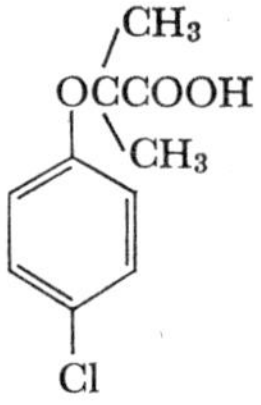

4-chlorophenoxyisobutyric acid (PCIB)

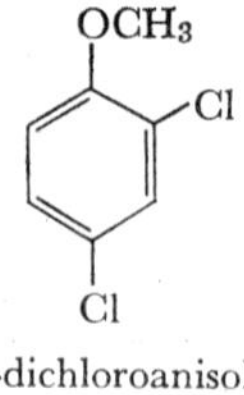

OCH_3, Cl, Cl

2,4-dichloroanisole

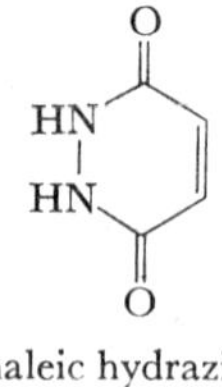

O, HN, HN, O

maleic hydrazide

Fig. 5.3. Anti-auxins

be prevented by supplying auxin at the same time, when presumably many more active auxin molecules reach the common site. But some anti-auxins do not act in this way. Tri-iodobenzoic acid (TIBA), for example, interferes with transport of endogenous auxin whereas others (e.g. ethylene) lower the native auxin content.

Ethylene is clearly chemically very unlike the auxins. It has many important effects on plants which include the promotion of leaf abscission and, in the form of ethylene chlorohydrin, the breaking of bud and seed dormancy. Ethylene, unlike the other anti-auxins, is a natural product, for example, of senescing leaves and ripening fruits. Since it has such important effects on plant growth it might well qualify as a naturally-occurring hormone, even though it is a gas.

The growth retardants

Some chemicals which were developed and used in the horticultural industry as growth retardants (Plates 6 and 7) have now become important in studies on plant growth and development. These substances include complex chemicals which are referred to as Phosphon, AMO, and CCC (chloro choline chloride).

When applied to plants these compounds appear to be acting antagonistically to the gibberellins; indeed, they are sometimes called anti-gibberellins. Furthermore, their action can be prevented by the addition of gibberellin. It now seems likely that the anti-gibberellins interfere with the biosynthesis of gibberellins in plants. This has been confirmed experimentally with cultures of *Fusarium moniliforme* which fail to synthesise gibberellins in the presence of, say, CCC, although otherwise growing and metabolising normally. Certain higher plant tissues (embryos of barley and wild oats) also respond similarly to CCC. These substances are being used increasingly by plant physiologists in studies of the roles of endogenous gibberellins (see, for example, Chapter 6).

Summary

A number of chemicals which are not synthesised by plants have the same physiological action as the auxins. Many of these substances are not so readily subject to enzymic regulation as the naturally-occurring auxins and they are therefore often used instead of the latter in horticulture and agriculture, especially as herbicides (i.e. at relatively high concentrations), to induce rooting, or to control fruit growth. Other growth regulators act as anti-auxins or anti-gibberellins. The latter, sometimes used as growth retardants, apparently inhibit gibberellin biosynthesis.

All these substances are frequently used in both the practical aspects of growth control and for investigations into mechanisms of plant growth.

6

Control of growth by hormones

Table 1 (page 46) lists the actions of the phytohormones in plant growth and development. The range of activities is very wide but in this chapter we will restrict our discussion to growth control and leave the effects of hormones on such processes as dormancy, flowering and leaf abscission until later chapters. As far as growth is concerned we may summarise the information in the table by saying that auxin primarily controls cell enlargement, gibberellin regulates both enlargement and division, the cytokinins predominantly affect cell division, and the inhibitors can interfere with the action of any of the other three hormones. (In some cases, though, auxin does stimulate division, e.g. in root initiation, and cytokinins cause expansion, e.g. in leaves.)

In one important way this summary is misleading because it suggests that the hormones act alone. Indeed, this view was held until fairly recently but now it has been abandoned, for it is becoming increasingly clear that the phytohormones interact with each other. The control of any growth process can no longer be attributed to any single hormone; instead, control is exerted by some or all of the phytohormones working together. For example, it seems that gibberellin cannot increase cell enlargement unless auxin is also acting at the same time. We shall discuss later some details of the interactions between hormones but it is necessary first to survey separately the growth-controlling activities of these substances.

Control by auxins

Studies on the growth and curvature of decapitated oat coleoptiles established that applied IAA influences cell growth. It is extremely potent and as little as 0·0025 μg ($2{\cdot}5 \times 10^{-9}$ g) is

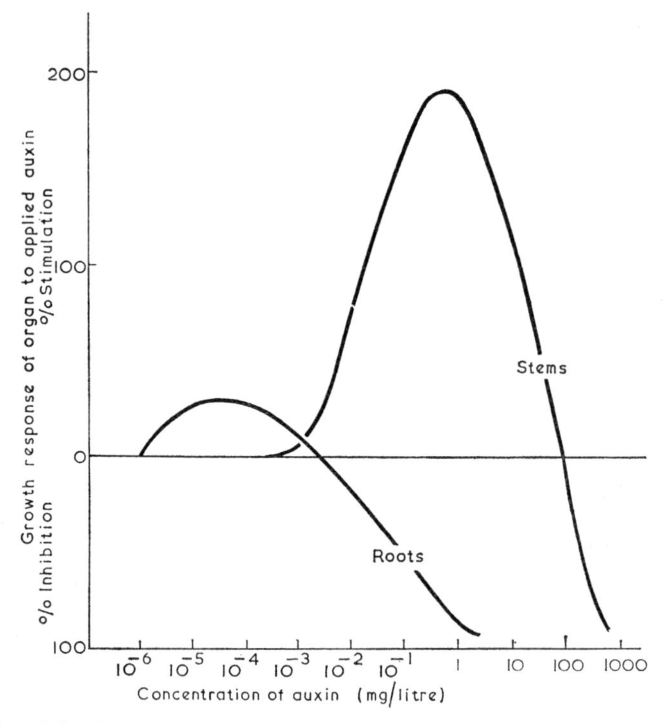

Fig. 6.1. The response curves of stems (and coleoptiles) and roots to applied auxin. (after Audus)

sufficient to induce a distinct curvature. Since an auxin (possibly IAA) is present in the tip of oat coleoptiles it seems reasonable to assume that it is this which controls growth of the elongating cells of the coleoptile. Evidence which is quite consistent with this view is that the growth which stops when the tip is removed is restored either by replacing the tip or by an agar block containing auxin. A similar situation is found in certain other organs. The flower stalks of the daisy, for example, stop growing when the flowers themselves are removed but growth is resumed when auxin is added in their place. Further, the growth of coleoptile or stem sections, (e.g. pea) initially measuring one or a few centimetres, is very much greater in solutions of IAA than in water alone.

Typical curves for the response of coleoptiles and roots to applied auxin are shown in figure 6.1. Concentrations of auxin as low as 1 part per million (approximately 10^{-5} M) greatly enhance coleoptile growth, but in roots even lower concentrations are effective. Both organs are inhibited by auxin concentrations above a certain value but clearly roots are much more sensitive than coleoptiles (and stems). The concentrations are, of course, the external ones, and we may expect that internally the values are considerably below these. The inhibitory parts of these curves were once thought to be significant so far as the kinetics of auxin action were concerned, but now it seems that the inhibition is brought about by ethylene, whose synthesis in the tissues occurs at high auxin concentrations.

Not all parts of the plant respond to added auxin. Overall growth of leaves or leaf discs is generally not stimulated, but only that of the veins so that distorted shapes arise. It is important to note in this context that growth of *intact* shoots and coleoptiles is usually not enhanced by externally applied auxin (cf. the gibberellins).

Do endogenous auxins control growth?

Stems, roots and leaves. The effects of applied auxin and the fact that auxins are found in plants together strongly suggest that growth control may depend significantly on endogenous auxin, but we would not yet be justified in concluding this. Much more direct evidence for this view derives from experiments and surveys which relate auxin content to growth. In pea stems, for example, a decreasing level of auxin accompanies the falling gradient of growth from the apical region of the plant to the base. More auxin is present in the buds of *Gingko* which are destined to become long shoots than in those which form short shoots; and in the apple tree seasonal changes in growth rate are associated with changing levels of auxin. Another relationship, but of a slightly different kind, is found in species in which the less rapidly growing tissues have a high IAA oxidase activity, this keeping the IAA level lower than optimal (e.g. in rosette stems of *Silene*). It is often assumed that root growth is negatively related to auxin content because the latter is at an inhibitory concentration.

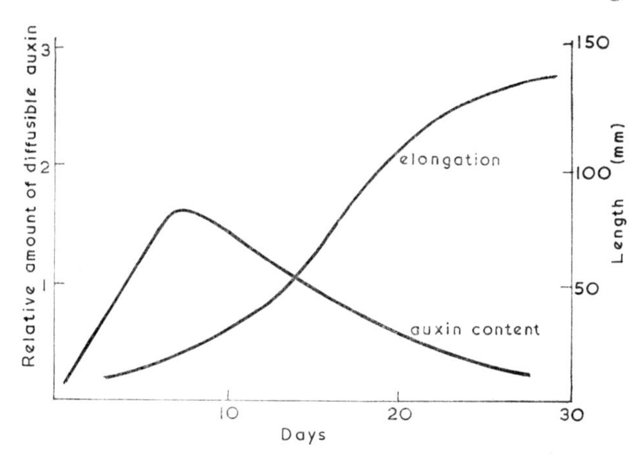

Fig. 6.2. The change in auxin content of leaves of *Solidago sempervirens* during their growth. Note that the maximum auxin content is found several days before the phase of maximum growth

Some experiments indeed enhanced cell elongation in the root by removing the apex (the site of auxin production) but this has rarely been repeated. On the other hand, in some roots (e.g. *Lens*) the highest auxin concentration occurs in the region of greatest growth.

Comparisons between growth rate and auxin content might not necessarily provide an adequate picture of the part played by auxin. If, for example, the rates of auxin production and utilisation (in growth) are both high, the level of extractable auxin is then low, and thus a high growth rate is accompanied by a low auxin concentration. Clearly in such a situation the auxin level at any one time is of little significance. In fact, there is often no obvious connection between the amount of auxin in a tissue and its growth. The auxin in leaves of *Solidago*, for example, continues to build up with no apparent relationship to the phase of maximum growth (figure 6.2). The explanation for some discrepancies like this might well be found in the theoretical case just presented, i.e. that auxin level is a result of synthesis and utilisation and the rates of these processes may not keep in step.

Fruits. So far most of our evidence for the control of growth by endogenous auxin comes from stems and roots; but a further striking example of the participation of auxins in growth is seen in the fruit. In fruit set the ovary is retained on the receptacle (i.e. abscission does not occur) and it begins to swell. Many years ago it was found that swelling of orchid ovaries was caused by substances in the pollen and that even extracts of foreign pollen were effective. Subsequently, it was found that auxin in pollen was responsible. Initial ovary growth, then, does not necessarily depend upon fertilisation but is induced by auxin diffusing from the pollen. Indeed, growth of an unfertilised ovary can be achieved by applying auxins (e.g. IAA, indolepropionic acid) either in solution or in lanolin paste. Such treatments cause not only initial growth of the ovaries of many plants, but this may even continue to produce ostensibly normal, though seedless fruit. This condition, called parthenocarpy, can be induced in tomato, squash and peppers, for example. Fruits such as banana, pineapple and varieties of oranges and grapes are naturally parthenocarpic and, significantly, they have an unusually high level of endogenous auxin (figure 6.3).

In most cases, though, ovary growth does not continue without true fertilisation. This is probably because the ovary tissues are incapable of making much auxin (except, of course, in the above seedless fruits) but must rely upon the hormone contributed by the fertilised ovules (i.e. the seeds). Work done with strawberries has demonstrated forcefully that seeds can control the growth of the flesh. The fleshy part of the strawberry is really the receptacle, and this bears the true fruits (pips) which for physiological purposes may be considered as seeds. Since these 'seeds' are easily removed or replaced the strawberry is particularly suitable for experimental studies. Growth of the receptacle in fact only occurs where 'seeds' are present, so that by selective removal of the latter from young strawberries various strange shapes result (figure 6.4). Further, since the action of the 'seeds' is simulated by auxin applied in lanolin paste even large 'seedless' strawberries can be formed. Significant quantities of auxin can be extracted from the strawberry 'seeds' and so it seems likely that this accounts for their control of flesh growth. This appears to be true also in

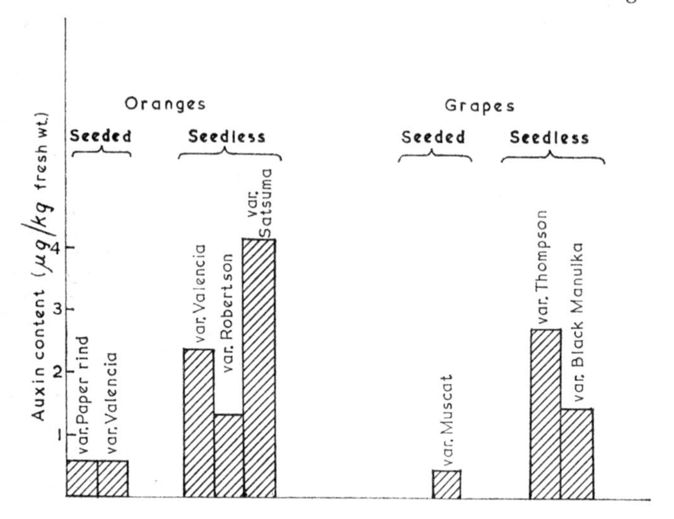

Fig. 6.3. The auxin content of some seeded and seedless fruit. The latter are, of course, naturally parthenocarpic, and have significantly high auxin levels

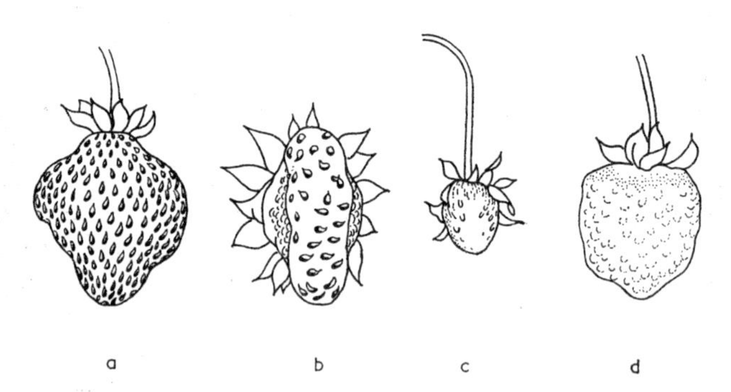

Fig. 6.4. The control of flesh growth in the strawberry. A normal strawberry with the 'seeds' left on is shown (*a*). When some of the 'seeds' are removed growth of the flesh occurs only under the 'seeds' which remain (*b*). Growth is stopped when all 'seeds' are removed (*c*) but is restored when the seedless strawberry is treated with auxins (*d*).

the apple where lop-sided 'fruits' will often be found to have seeds only on one side – that which has grown. But the control of all fruit growth does not lie only with the auxins. Since only a minority of species can be induced by auxins to form parthenocarpic fruit other factors must be involved. This is also indicated by those fruits which have a double sigmoid growth curve, where frequently the highest auxin concentration occurs in the rest phase, and the second growth increment seems unrelated to auxin level. We shall see later that gibberellins and cytokinins are certainly involved in fruit growth.

There is considerable evidence, therefore, to suggest that auxin participates in the internal growth control mechanisms in various organs. We will now go on to see how gibberellins are involved in plant growth.

Control by gibberellins

Action of applied gibberellins

Stems and hypocotyls. Some of the most dramatic effects of gibberellins are on stem growth of intact plants. This is perhaps not unexpected since these chemicals were discovered as a result of the effect of the pathogenic fungus on stem growth of rice.

When gibberellic acid is applied as a small drop of solution to the leaf blade or axil, stem growth of many species is very greatly increased. Cabbages grow several feet tall and lettuces become vine-like, growing around a central support (Plate 7). Dwarf plants also respond in a spectacular manner, and their growth is increased so much that they become indistinguishable from the normal, tall varieties (Plate 7). Many dwarfs (e.g. dwarf maize, peas) are mutants of wild type, tall varieties, and so gibberellin therefore causes a phenotypic reversion of a genetic character, although the progeny of such plants are, of course, still dwarfed. The treated plant has a long stem with little or no branching although the untreated dwarf may have a bushy habit (e.g. dwarf bean). Thus, gibberellin also enhances apical dominance (see page 95).

In some plants, for example the garden pea, the increased

overall growth of the stem is attributable entirely to great internodal elongation as the total number of internodes is unchanged. The action of the gibberellin is largely restricted to those internodes still capable of elongating when the chemical is applied. Other species, however, show an increase in length and number of internodes (e.g. dwarf French bean, sweet pea) and so in these cases gibberellin must also increase cell division at the apex. In the cereals (e.g. maize and rice) internodal growth, the extension of leaf sheath and blade are all promoted, and also the activity of the intercalary meristem is enhanced.

Applied gibberellins also stimulate great enlongation of seedling hypocotyls. In lettuce and lentil, for example, the chemical increases both cell size and number. Hypocotyls are often used in bioassays of the gibberellins as an alternative to dwarf plants (figure 6.5).

The response of both stems and hypocotyls to gibberellins depends on the presence or absence of light. Many dwarf and normal varieties show equal excessive growth in darkness and both then exhibit little or no response to gibberellic acid. Light, however, strongly inhibits growth of the dwarfs more than that of the talls but the inhibition is relieved by GA_3 (see page 83 for further discussion). Similarly, the promotive action of GA_3 on lettuce hypocotyl growth is evident only in the light since they grow tall in darkness even without the hormone.

Another remarkable action of gibberellins on stem growth is found in certain rosette plants. In these plants the nodes are so tightly compacted that all the leaves arise very close to each other apparently from the top of the root. When treated with gibberellin many such plants (e.g. *Hyoscyamus* spp, carrot) rapidly elongate due to stimulation of cell division, at the apex and elsewhere, and cell elongation (Plate 8). Without gibberellin this 'bolting' is normally induced by exposure of the rosette to low temperatures or long days (depending on the species), and is accompanied or followed shortly by flower development on the stems.

An important point which should be emphasised is that all the above effects on stem growth are obtained by applying gibberellins to whole, intact plants. This sharply contrasts with the case of the auxins where responses are only rarely elicited under such conditions. This is possibly because growth of intact plants is not generally auxin-limited but might well be limited by the normal level of endogenous gibberellins; or, put in another

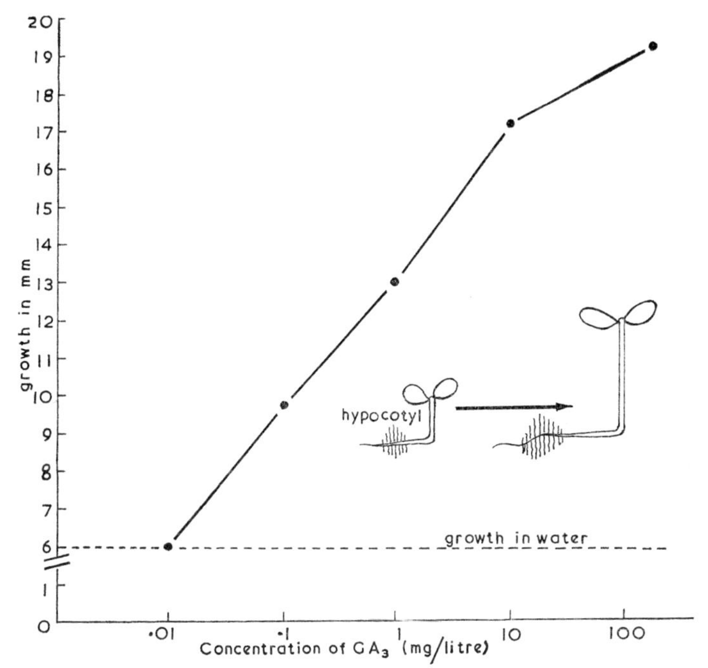

Fig. 6.5. The response curve of lettuce hypocotyls to added gibberellic acid. The gibberellic acid solution is applied to the roots of the seedlings which are growing on filter paper

way, intact plants are not normally deficient in auxin but are in gibberellin.

The different gibberellins do not affect growth equally. Dwarf peas are extremely responsive to GA_3, about one-third as sensitive to GA_7, but only one-twentieth as sensitive to GA_4. In contrast, lettuce hypocotyls respond more to GA_7 than to GA_3, while cucumber hypocotyls are almost completely unaffected by GA_3 at comparable concentrations. These differences have been thought to be due to the inability of certain species to convert the applied gibberellin into an active form, rather like the inability of many plants to change IAN to IAA. This is somewhat doubtful, however, as a particular gibberellin may be inactive on stem growth but will induce flowering of the same plant.

Roots, leaves, fruits and seeds. Roots of very young seedlings (e.g. lettuce) grow more when supplied with certain gibberellins, but generally roots of intact adult plants show little response. However, excised roots growing in tissue culture are promoted by the presence of gibberellic acid in the liquid medium.

Though inactive in many plants the gibberellins do increase leaf growth (i.e. by cell elongation) of some species (dwarf bean, sweet pea, cereals). Leaf growth even in darkness is often enhanced although normally light is required for the promotion of expansion. Here, then, these hormones are simulating the action of light, a very important property which we shall consider in more detail during our discussion of the environmental control of growth and development. Changes in leaf shape which are normally associated with juvenility and maturity are also induced by gibberellins. *Eucalyptus* leaves are transformed from a typical juvenile to a mature shape whereas in ivy the effect is the reverse.

The gibberellins are often more effective than the auxins in promoting parthenocarpy, for example in the tomato. Some species such as cherry, apricot, peach and the rose, *Rosa arvensis*, are completely insensitive to auxin but readily set fruit after treatment with gibberellin.

Finally, applied gibberellins have an important and interesting action quite different from that of auxins; this is their ability to cause resumption of growth in dormant seeds and buds. We will return to this in the appropriate chapter, but we might note here that both cell division and elongation are promoted.

Do endogenous gibberellins control growth?

In view of the many dramatic effects of applied gibberellins on growth it seems reasonable to suggest that endogenous gibberellins have similar actions. But to be sure of this we must find further, more direct evidence.

Dwarf plants, which respond so dramatically to gibberellins, would appear to be suitable objects in which to look for such evidence, for the difference between dwarf and tall varieties might possibly reflect their respective gibberellin contents. Unfortunately, this does not generally seem to be the case, although it apparently is so in the two varieties of Japanese morning glory (*Pharbitis*). There is, however, another possible explanation for dwarfism. The dwarf habit depends upon light, as the difference between dwarf and tall varieties is lost when the plants are grown in darkness. Now, two gibberellins, probably GA_1 and GA_5, have been extracted from dwarf peas and in darkness both are effective on dwarf and tall peas. In the light, however, GA_5 is still effective on the tall type but very much less on the dwarf variety. Light therefore apparently prevents the action of one of the endogenous gibberellins of the dwarf pea. This very interesting demonstration of the action of gibberellins on the same species from which they were extracted provides good evidence that endogenous gibberellins are involved in growth control.

Further evidence comes from some rosette plants. As we have already mentioned, these bolt in response either to gibberellin treatments or to the environmental factors of daylength and low temperature. It has been shown that the level of extractable gibberellins increases after the appropriate daylength treatment. Evidently, bolting is caused by this increase in endogenous hormone.

The chemicals CCC, Phosphon and AMO (Chapter 5) prevent gibberellin synthesis in *Fusarium* or in certain higher plants. Since these substances also retard the growth of higher plants it is plausible to suggest that normal growth requires continued gibberellin biosynthesis. Cytological studies demonstrate that the growth retardants act at least on the apex where cell division is hindered. When gibberellic acid is added, the mitotic activity and the growth are restored to normal (figure 6.6). This, and similar examples, strongly support the view that endogenous gibberellins act as growth-controlling hormones.

We have now seen that two phytohormones – auxins and gibberellins – are concerned with growth control. How are the other two kinds of hormone involved?

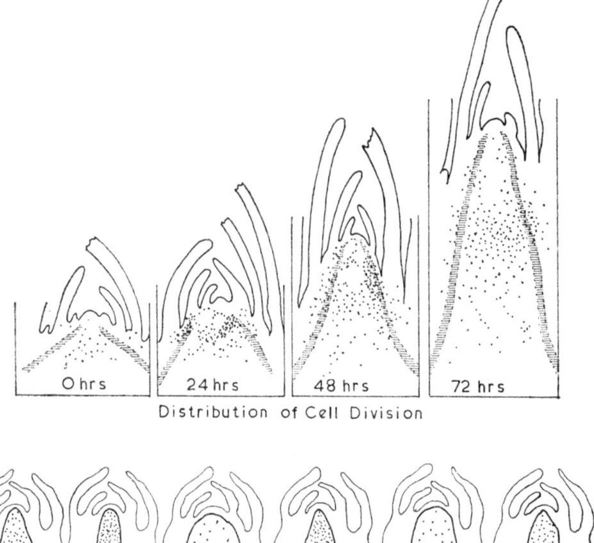

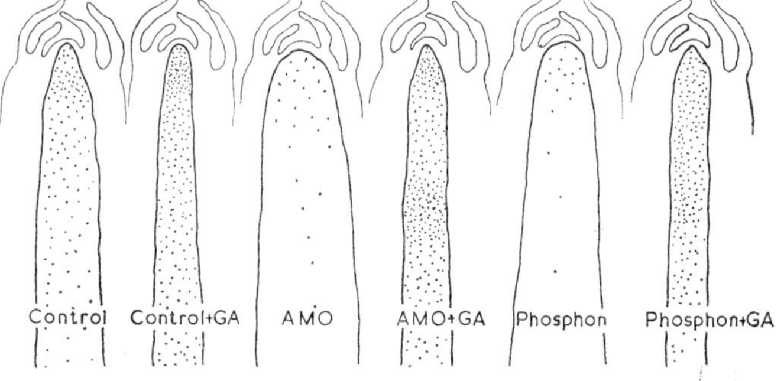

Fig. 6.6. The effect of GA and growth retardants on mitotic activity.
Upper figures: These show the increase in mitosis in a rosette plant (*Samolus*) following application of GA. Each dot represents one cell division. Note the increase in frequency of divisions in the sub-apical meristem (the hatched area is the vascular tissue)
Lower figures: These show the mitotic activity in the sub-apical meristem of *Chrysanthemum*, treated with GA, with growth retardants, or the two together. Note how GA restores the mitotic activity of plants treated with growth retardant to that of the control. (after Sachs)

Control by cytokinins

The effect of applied cytokinin

The action of cytokinins on plant growth is quite different from that of auxins and gibberellins and is also more limited. They affect overall growth of a very few plants when absorbed by the roots. In duckweed, kinetin stimulates frond production especially in darkness. Parthenocarpy is induced in some species, as is germination of certain seeds, breaking of dormancy and the outgrowth of axillary buds (see page 96). In most of these cases the action of the cytokinin is basically on cell division, which is not surprising since this is how they were discovered (Chapter 4). Because of their effect on cell division they also influence differentiation and morphogenesis. We will consider this later in Chapter 10 but we might note now that applied cytokinins promote lateral root formation, and bud formation on calluses, roots (e.g. *Isatis*), leaf cuttings (e.g. African violet) and on moss protonemata. Most of these organs already have an innate tendency to form buds which seems to be greatly enhanced by the cytokinin. In some of these effects, however, cytokinin alone is not so active but the addition of auxin is also required.

One instance involving only cell elongation is when kinetin enhances leaf expansion in darkness. In this case kinetin simulates the action of light (Chapter 9).

The role of endogenous cytokinin

The facts that applied cytokinins have such profound effects on plants and that these substances are also found within plant tissues make it very likely that they are implicated in the systems of growth control. Nevertheless, the role of endogenous cytokinins is not as clear as that of auxins and gibberellins, and only a few examples are known in which internal cytokinin levels seems to be related to growth activity. For example, the cytokinins in young apple and plum fruitlets are at their highest level at the time of intensive cell division and they decline just before this stops. In fact, many actively dividing tissues frequently contain more cytokinin. Apart from relationships between cytokinin and growth, others are known, for example in senescence (Chapter 13).

Control by inhibitors

Inhibitors in plants play an important part in certain dormancy phenomena where, for example, they may cause the cessation of shoot growth. Extracts which inhibit growth (say, of oat coleoptiles) have been obtained from various plants whose growth is suspended due to certain environmental factors (Chapter 12). One of the inhibitory substances is abscisic acid which is one of the very few to have been investigated in detail (page 167). This inhibitor also decreases IAA-inducible coleoptile growth but the effect is relieved by gibberellin. Another inhibitor, still chemically unidentified, which interacts with gibberellin, occurs in lima beans. This inhibitor reduces the action of GA_3 on dwarf peas, and presumably also interferes with gibberellin action in the bean itself.

Some phenolic substances, such as p–coumaric acid in pineapple, depress growth by promoting the oxidation of IAA, probably by activating IAA oxidase. Other phenolics (e.g. chlorogenic acid in rice coleoptiles) inhibit growth at certain concentrations but promote IAA-dependent growth at lower concentrations, a dual action which is difficult to explain.

In this discussion on inhibitors we have been particularly concerned with interactions between these substances and other hormones. This is logical since where inhibitors and promoters are present together one would expect that the final growth of the plant would be a resultant of their different actions. But the same might also be said for the promoters themselves. Let us now consider how these work together to control plant growth.

Interactions between hormones

To what extent do the auxins, gibberellins and cytokinins interact with each other? Does the action of one hormone depend on the presence and activity of another? How is overall growth

Plate 7 (*a*) *The effects of gibberellic acid on growth of lettuce.*
The tall plant has been treated with a few micrograms of the plant hormone, which has caused excessive stem elongation and flowering.

(*b*) *The effects of gibberellic acid on growth of dwarf peas.*
Gibberellic acid (in alcoholic solution) was applied to the first true leaf of the seedlings; these were photographed 26 days later. (From right to left; untreated; alcohol control; 0·01 μg GA_3; subsequently the dose increased in twofold steps to the highest at 10.24 μg GA_3).
Note that very small amounts of GA_3 (0.1 and 0.2 μg) cause a significant increase in height, and that there is possibly only a slight inhibitory action of the highest doses (cf. the auxins).

Plate 8 *The effects of gibberellic acid and low-temperature treatment on flowering of carrot* (Daucus carota).
The plants on the left and in the centre are first-season plants; that on the left has not been treated and remains as a typical non-flowering rosette. The plant in the centre has had a daily application of 10 μg GA_3. This plant has bolted and formed flowers and appears identical to the second-season plant on the right, which has received the cold treatment normally necessary to induce flowers in this biennial.

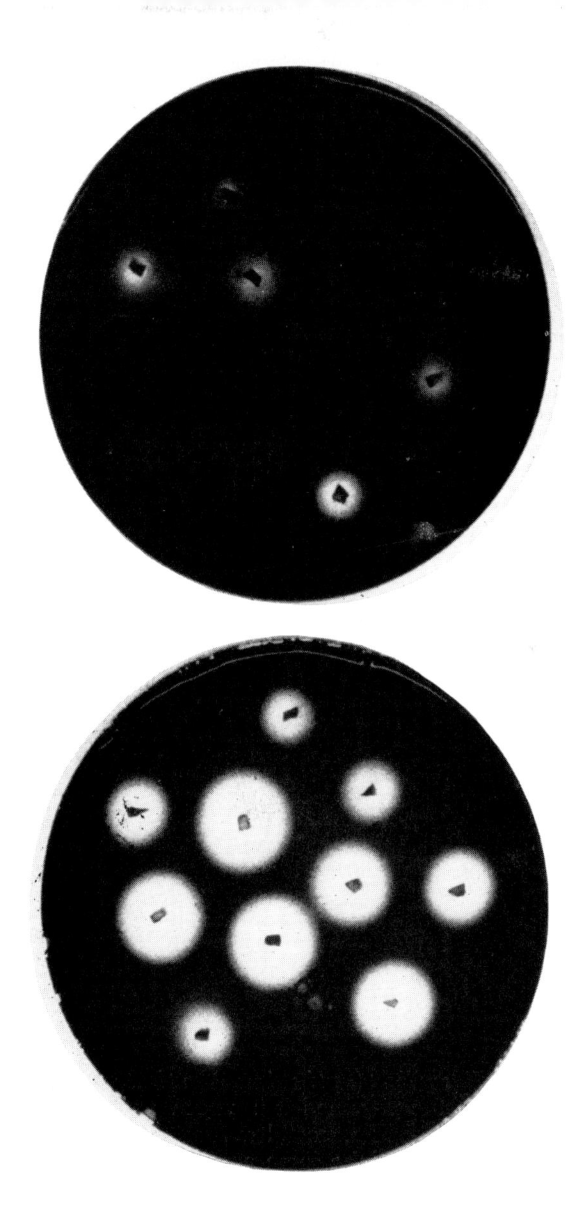

Plate 9 *Induction of α-amylase in barley aleurone by gibberellic acid.* The clear areas around the aleurone pieces (10 in each treatment) indicate the presence of α-amylase which has been released from the cells and broken down the starch. Note that significant formation of the enzyme occurs only when the aleurone tissue is in the presence of GA_3.

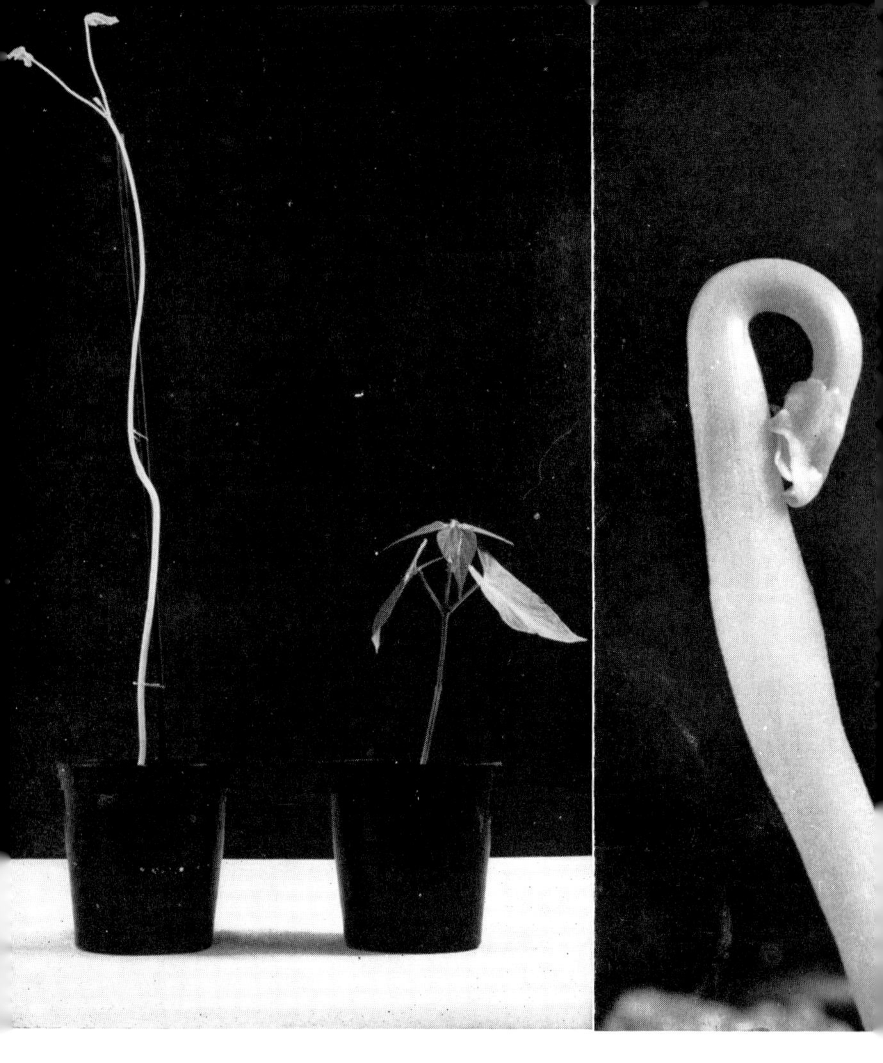

Plate 10 *The effects of light (and its absence) on plant growth.* The bean plant on the extreme left has been grown in darkness, and has very long internodes, small leaves and a pale colour. The accompanying plant is of the same age and was grown in the light. The photograph on the extreme right is of the apical bud of a bean seedling growing in darkness. Note the plumular hook and the small leaves. On receiving light, the hook straightens.

controlled by the total hormone content? These are some of the questions which we must now try to answer.

The dramatic effect of gibberellins on stem growth certainly depends on the presence of auxin. When the internal source of auxin, the apex, is removed, internodes which are normally induced by gibberellin to elongate three or fourfold fail to respond. The response is almost completely restored when auxin in lanolin paste is applied in place of the apex. A similar dependence on auxin for gibberellin-controlled growth can be demonstrated with stem sections. The growth of pea-stem sections which are floated on solutions containing gibberellic acid alone is virtually no greater than when they are in water. But when a small amount of IAA is added the gibberellic acid then causes an appreciable increase in growth. Similarly, GA_3 does not induce a growth curvature when applied to one side of the top of a decapitated coleoptile. When IAA is applied at the same time the resulting curvature is greater than if the IAA were alone. Clearly, GA_3 enhances auxin-induced growth curvature.

Both GA_3 and auxin are in fact required for elongation not only of stems, but also of petioles, leaves and hypocotyls. Growth of the hypocotyl of intact cucumber seedling is promoted by applying IAA or GA_3. AMO, which inhibits gibberellin biosynthesis (page 71) reduces the response to applied auxin, and an anti-auxin hinders the action of applied GA_3 (figure 6.7). Evidently GA_3 can promote hypocotyl growth only when the endogenous auxin is functioning, and applied IAA can act only if continued synthesis of GA in the hypocotyl is taking place.

Transverse growth (increase in girth) of stems is best promoted by auxins and GA_3 supplied together, and similarly both substances are apparently required for stimulation of division and differentiation of the secondary cambium. In these cases where cell division is occurring the endogenous cytokinins are also likely to be involved. Growth of some organs, such as excised roots, is in fact more responsive to a mixture of auxins, gibberellins and cytokinins than to any one supplied alone. Similarly, fruit growth is probably controlled by the three hormones acting in unison or at different growth phases. Since growth of this

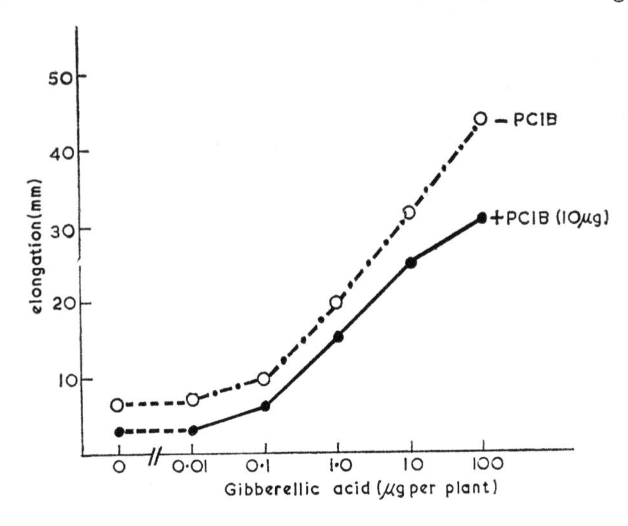

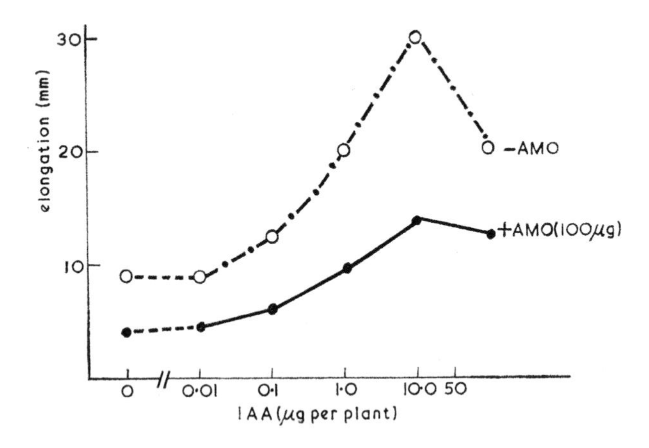

Fig. 6.7. The affect of GA_3 and IAA on the growth of cucumber hypocotyl in the presence or absence of anti-auxin (chlorophenoxyisobutyric acid, PCIB) or growth retardant (i.e. the 'anti-gibberellin', AMO). Note how PCIB reduces the action of GA_3 (upper graph) and AMO reduces the action of IAA (lower graph). (after Katsumi, Phinney and Purves)

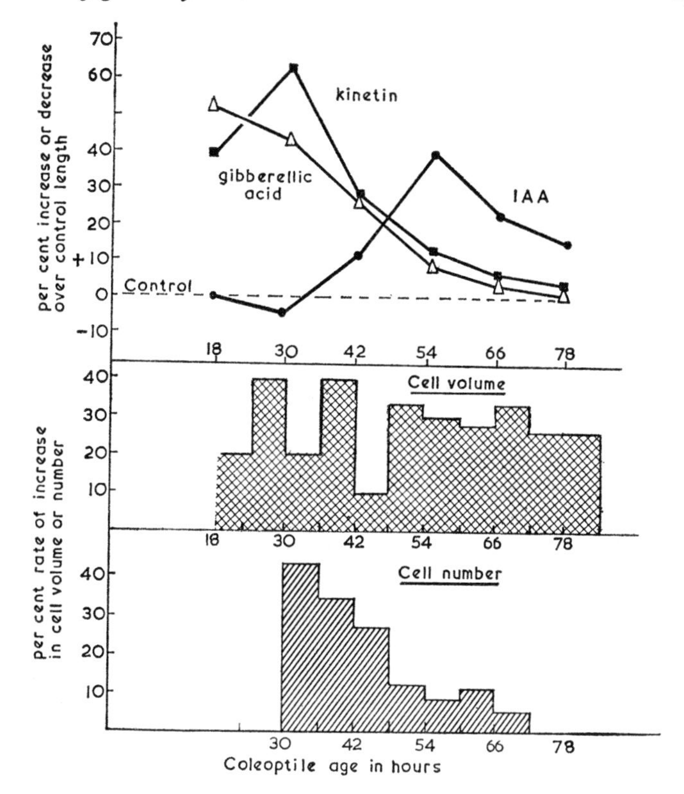

Fig. 6.8. The changing growth response of wheat coleoptiles to added hormones related to the pattern of cell division and enlargement. Note that kinetin is active at the cell division phase, when IAA is inactive. GA_3 seems to act on the earlier phase of elongation (or division), and IAA on the later phase of elongation (after Wright)

organ involves cell division (especially when young) and elongation, we would not be surprised to find all three hormones concerned. The reason why some fruits are more sensitive to applied auxin, some to GA, and some to cytokinins may simply reflect the different natural levels of one or other of these hormones within the tissues. For example, if the ovary is low in gibberellin but not in auxin and cytokinin then application of GA, but not auxin, will promote growth.

Some plant tissues, as they age, change in sensitivity to the hormones. Wheat coleoptiles, for example, respond to GA_3 and to kinetin when young but little to auxin; response to auxin increases with age (figure 6.8). Findings like these support the idea that the complete cycle of growth may be governed by the hormones acting in sequence. As far as stems are concerned, the production of new cells in the apex might be controlled predominantly by gibberellins and cytokinins. These cells become older and are pushed further back from the apex, when they begin to enlarge. The enlargement itself is regulated by auxins and gibberellins. The growth of other organs may be similarly controlled.

Thus, we are beginning to form a picture of the internal mechanisms of growth control. Knowledge of the phytohormones has clearly developed enormously since the early studies on coleoptiles. At least four kinds of hormone are involved in the regulation of growth, which depends on the proper interplay among them. But even such a view of plant growth is too simple, and we cannot yet fully explain growth control in terms of the known hormones. Perhaps more exist; but certainly many other factors – sugars, metal ions, vitamins, lipids – are also involved. We must also remember the environmental factors of light, temperature, gravity, etc. which can effect the internal controls and which ultimately call the tune. We are only just beginning to understand the manner in which all of these interact to result in the regulated and co-ordinated events which comprise plant growth.

Summary

Auxins affect cell enlargement when applied to stem or coleoptile sections but at the same concentrations have little or no effect on intact organs. There is good evidence, however, that endogenous auxins control cell expansion in stems and other organs (e.g. fruits). Gibberellins, too, greatly stimulate cell enlargement and do so even when applied to intact plants. Moreover, they can also promote cell division at the apex and elsewhere. There is convincing support for the view that endogenous gibberellins are involved in growth control in stems, leaves, fruits and seeds, and roots. Endogenous cytokinins must also be involved in growth processes, particularly in cell division, but occasionally in enlargement (e.g. leaves) but their role is not so clear. Hormonal control of growth is not ascribed to a single hormone, then, but depends on the interaction between, at least, the four categories discussed in this chapter.

7

Co-ordination and correlation

The parts of the plant are co-ordinated, i.e. they influence each other so that growth and development of an organ or tissue may be determined by the activities of another. Of course, the leaves, via photosynthesis, ultimately supply the necessary bulk for the growth of all the plant, and the roots similarly are essential as organs which absorb water. But the term correlation, which is used specifically to describe the influence of one part of the plant over the growth and development of another has a special connotation. Direct nutritive effects of these kinds are usually omitted. In correlation, the hormones seem generally to be involved.

We have already touched upon some of the important interrelationships between plant parts which have a hormonal basis. The stem or coleoptile apex is the source of those auxins which control cell elongation in the tissues lower down. In the tropisms, differential growth is instigated by root and stem apices which perceive the stimuli. Cytokinins and gibberellins synthesised in the roots travel upwards in the xylem to the shoot, and cellular division and elongation of stem and leaves may well rely upon this contribution. Furthermore, we have seen that growth of fruit flesh can be regulated by the seeds which produce auxins and probably also gibberellins.

A rather interesting example of how the different regions of the plant interact to regulate development is found in the potato (figure 7.1). In this plant, lateral shoots grow out of the lower leaf axils to form underground stolons which swell at their tips into tubers. However, if the main apex of the shoot is removed the stolons become erect, develop leaves, and do not form tubers. Clearly, this is a form of apical dominance (see below) and in

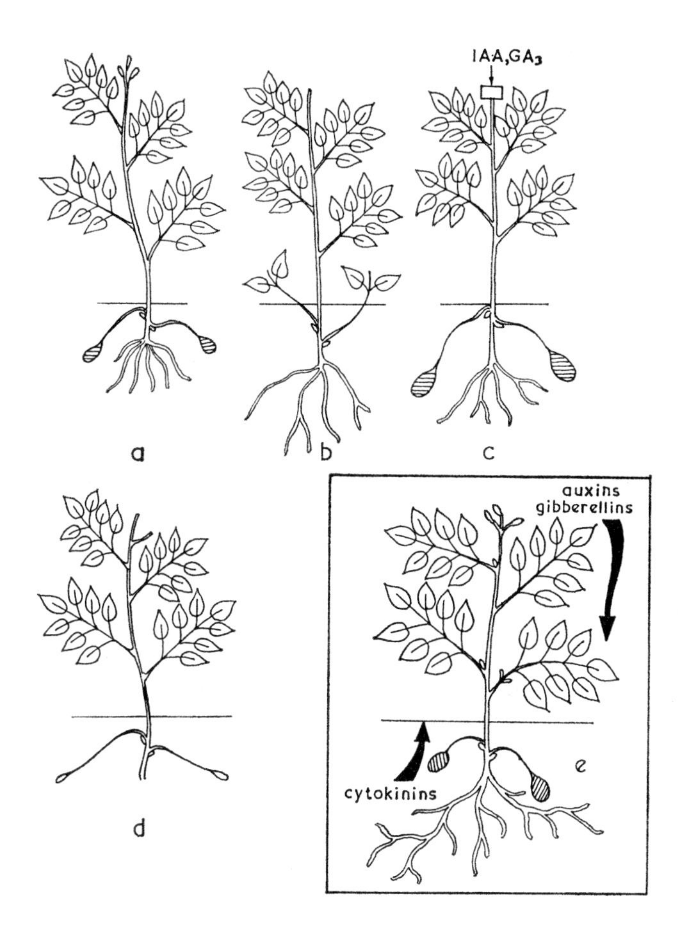

Fig. 7.1. Stolon formation in the potato plant. This process is one aspect of apical dominance in this species. *a* = normal intact plant. *b* = decapitated plant. *c* = decapitated plant with auxin and gibberellin added. *d* = decapitated rootless plant. The behaviour of the stoloniferous shoots is controlled by auxins and gibberellins from the apex and cytokinins from the roots – *e*.

this case it is caused by both auxins and gibberellins, for if these two hormones are applied to the decapitated stem normal stolon formation still proceeds. Cytokinins are also implicated since a stoloniferous shoot can be made to turn up and grow erect by treatment with these hormones. Now it is interesting that decapitation of the plant induces upward growth of the stolons only if roots are present and, as we have already seen, the roots are sources of endogenous cytokinins. It seems, therefore, that the formation of stolons is determined by an interaction between the stem apices and the roots, or, more accurately, between the auxins, gibberellins and cytokinins which they contain.

Cambial activity

In many species, renewed cambial activity in the spring, producing the secondary xylem and phloem, is induced by near-by buds emerging from dormancy. After initiation, continued cambial activity can often be shown to depend upon adjacent leaves. The leaves exert a similar influence in herbaceous plants. Auxins and gibberellins are almost certainly involved in these effects. These chemicals, added together in place of a bud, stimulate division and subsequent differentiation of cambial cells and it is well known that buds contain these hormones.

Root initiation

Buds and leaves promote the formation of roots on stem cuttings. Cuttings of many species form roots spontaneously but this can be hindered by removal of the leaves, or in woody twigs, the buds. Rooting is strongly polarised and often occurs only at the very base of the stem (though sometimes each node produces roots). Evidently there is a downward flow of some 'influence' from the leaves or buds which initiates roots where it accumulates (e.g. at the cut end). Auxins are suspected to be causal agents here, a suspicion which is strengthened by their ability to induce root initiation when applied to cuttings.

Apical dominance

Some correlative actions involve the prevention of growth – this is called correlative inhibition. Perhaps the best-known example is apical dominance; we have seen one instance of this at work in the potato plant. Every gardener is familar with apical dominance which he neutralises when he removes the tip of a shoot to encourage branching. The intact apex prevents the growth of axillary buds lower down the stem, although in many plants only the lateral buds nearest the apex are inhibited. Apical dominance is, of course, lost when the shoot tip is removed, the buds grow out, and a branched plant results (figure 7.2). The phenomenon is apparent also in woody twigs bearing dormant buds. One aim

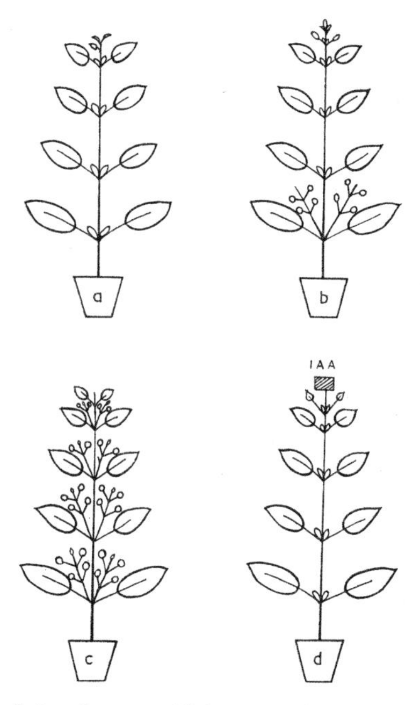

Fig. 7.2. Apical dominance. This can take several forms including complete suppression of lateral bud growth (*a*) or only partial suppression (*b*). Apical dominance is removed by decapitating the plant (*c*) but is restored by adding IAA (*d*).

in pruning some fruit trees is to encourage growth of the lateral buds by removing the terminal one. In lilac the terminal apex forms the inflorescence and while this is present the vegetative buds beneath this do not develop. As soon as the inflorescence dies, or is pulled off, bud growth commences. Lateral buds on tubers (e.g. potato) and lateral roots in some species are also subject to apical dominance.

In certain plants apical dominance is not expressed in the inhibition of bud growth but takes another form. Rates of growth of the lateral branches and the angles which they adopt in relation to the main stem are controlled by the main apex.

Many experiments implicate auxin in apical dominance. Even if the main apex is removed, lateral bud growth is prevented when auxin is applied in its place. The apex is a major site of auxin production and it has been suggested that this hormone, moving down in polar fashion, causes lateral bud inhibition. It is curious though, that auxin is supposed to stop growth of the lateral bud apices but permit that of the main apex which is, after all, anatomically similar. One possible explanation of this paradox is that lateral buds are sensitive to much lower auxin levels and that the hormone reaching them from the main apex is at the inhibitory part of the concentration curve.

Competition for nutrients has also been thought of as a cause of apical dominance. The main apex might attract food material which thus by-passes the lateral buds. Evidence for this view comes from the finding that apical dominance is absent in some plants (e.g. flax) under conditions of high nitrogen supply. Auxin might, nevertheless, play an important part by acting as the mobilising agent. Radioactive substances supplied to the leaves are known to be attracted to the apex or to auxin applied in its place. Another way in which auxin may act in apical dominance is to prevent DNA synthesis in meristematic cells of lateral bud apex. This has been shown to occur in *Tradescantia* but its general significance has not yet been confirmed.

Auxins, though, are not the only hormones involved in lateral bud inhibition. Inhibited lateral buds can be released by applying kinetin to them. In stem sections bearing one lateral bud, which are floated on solutions, the promotive action of kinetin is opposed

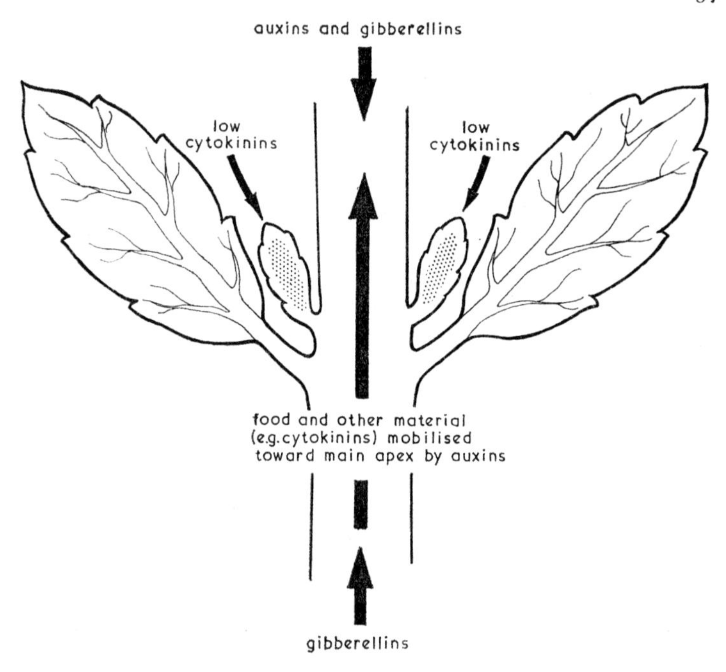

Fig. 7.3. Hormones and lateral bud inhibition. Auxins, gibberellins and cytokinins coming from different parts of the plant may be involved in the inhibition of lateral bud growth. Interference with the movement and internal concentrations of these hormones may profoundly modify the inhibition

by auxin in the solution. We should recall also, that gibberellin treatments frequently enhance apical dominance (page 79) and it has recently been noted that this hormone amplifies the action of auxin when the two are applied together to a decapitated plant. Thus, we may begin to construct a picture of apical dominance in terms of hormonal interactions (figure 7.3).

Effects of leaves

We have emphasised the part played by the apex in lateral bud inhibition and have interpreted this on a hormonal basis. However, in many plants the leaves, not the apex, are responsible for bud inhibition, and their effect can often be observed for some distance up and down the stem. Sometimes, both the leaves and the apex control lateral bud growth. In plants of this kind, lateral bud growth is poor even when the apex is removed, but they begin to grow out when their respective leaves are taken off. Leaves may also influence each other or various associated structures. For example, removal of some leaves often induces the remaining ones to expand. On the other hand, in the broad-bean plant, the stipules enlarge when the laminae of the young leaves are removed. We still do not understand the basis for these effects of leaves.

Abscission

The shedding of organs from the plant can also be considered as an aspect of correlation. We have already hinted at one example of this during our discussion of fruit set. Here, the abscission of the young ovary is possibly prevented by auxins produced by the developing seed. Perhaps the most obvious case of abscission during the life of the plant is leaf fall, which in most deciduous plants of temperate climates, takes place in autumn. Daylength has an influence since often abscission can be delayed by artificially lengthening the daily light period (this can be seen on branches close to street lamps).

Leaf fall occurs because the cell walls of the abscission layer at the base of the petiole apparently disintegrate. This event normally does not happen until the leaf blade begins to age (senesce) at the end of the growing season, but readily takes place even in a young petiole if the lamina is removed. Clearly, then, in a young leaf the formation of the abscission layer is prevented by the lamina. The effect of the lamina is reproduced if auxin is applied to the debladed petiole (figure 7.4). Since young leaves contain considerable quantities of auxins it seems evident that these are the agents which stop the disintegration of the abscission layer.

Auxins, however, do not operate in a straightforward manner in abscission nor are they the only important substances. The auxins might have at least a dual action, inhibiting abscission relatively early in the life of the leaf, but promoting it later. Abscission is stimulated by ethylene which is now known to be produced by many plant tissues including senescing leaves and fruits. Ethylene, then, might be responsible for inducing senescence of the lamina which precedes leaf fall. Further, senescent leaves do produce some other substances, not auxins, which can promote abscission.

Little is known of the detailed biochemical mechanism of abscission but there is evidence that in *Phaseolus* a high cellulase activity in the abscission zone is induced by the ageing lamina.

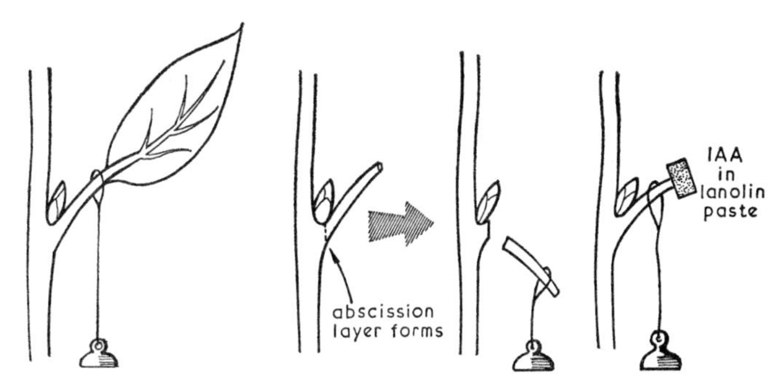

Fig. 7.4. Petiolar abscission is controlled by the lamina. If the lamina is removed an abscission layer quickly develops, and the petiole falls off. The development of this layer can be prevented by auxin

This enzyme breaks down the cell walls. Thus, the control exerted by auxins via senescence might ultimately be on enzyme activity or enzyme synthesis.

Leaf senescence can itself be influenced by other organs. Often, leaves begin to senesce only when the flowers have been fertilised and fruits are developing. The fruits, in fact, appear to be responsible. Once more, a gardening practice – this time the removal of developing fruits – owes part of its physiological basis to a correlative effect. Auxins, gibberellins and cytokinins are relatively abundant in young, growing fruits. These hormones, by acting as mobilising agents, draw off nutrients thus depleting the supply to the leaves. In fact, the attraction into the fruit of the cytokinins themselves, presumably from the roots, is thought to be involved in correlative aspects of leaf senescence. In Chapter 13 we will see how the cytokinins in the leaves are important in opposing the onset of senescence. This provides us with another example of how interaction between hormones is concerned in the regulation of the plant's activity.

Summary

We have now seen several ways in which different parts of the plant influence each other, e.g. in stolon formation, root-initiation, lateral bud inhibition. In many of these cases a satisfactory explanation can be given on the basis of the action and interaction of various hormones, but in some that we have dealt with this is not yet possible. Whatever the causes are it is evident that correlation is important in determining the direction of plant growth and development.

8

How might hormones work?

In the last few years important and exciting advances have been made in the study of hormone action but we are still not able to give a satisfactory account of how these chemicals work. A major difficulty is that all the hormones have so many different physiological effects, as we can see from the fact that auxins and gibberellins, for example, qualify for inclusion in almost every chapter of this book! It would be scientifically pleasing if we could uncover a basic mechanism to explain these multifarious effects but we are still a long way from this. Nevertheless, we shall see in this chapter how recent work has begun the move in this direction, for all the phytohormones are apparently involved in nucleic acid and protein synthesis, these being processes at the centre of the control of cellular activities.

The auxins, the longest-known phytohormones, have been most intensively investigated with regard to hormonal mechanisms. Their activity is presumably due to their chemical nature and so the early studies were concerned with seeing if the many different auxins (i.e. including the synthetic ones) had any underlying chemical similarity. Once these common chemical characteristics were clarified, it was thought, the cellular action of auxin might be more readily understood. From knowledge of the formulae of auxins a list of structural requirements for activity was, therefore, drawn up. It seemed that an aromatic ring was required and a side chain containing a carboxyl radical or one easily convertible into this radical. A certain spatial relationship between the ring and the side chain was apparently necessary and any substitution in the ring, say by chlorine, must leave a free ortho position. Other features, such as the length of the side chain, may influence transportability of the auxin. These characteristics seemed in the 1930's to be well documented, but nevertheless they afforded no insight into the mechanism of action of the auxins. Moreover, these structural requirements have recently been thrown into confusion by substances such as some carbamates, certain chelating agents and even ethylene, all of which may have some activity as auxins, but chemically resemble them very little. Discrepancies like these are not so evident in the case of the gibberellins (figure 4.6), although physiological activity has been suspected in some compounds, still unidentified, but which seem chemically dissimilar to the characterised hormones. Nevertheless, knowledge of the chemistry of the gibberellins has also not yet contributed to our understanding of their mechanism of action.

With some exceptions, most of the cytokinins are purines having varying degrees of complexity. This fact may be a highly significant clue and the cytokinins may, indeed, be the only hormones whose chemistry leads us immediately to suspect a particular biochemical role.

Almost all the researches into the mechanism of auxin action has been done on extension growth. At a physiological level the question was asked, 'What part of the process of cell growth might auxin affect?' Earlier (Chapter 3) we saw that cell growth involves increased entry of water. This may be promoted by changes in the factors concerned in the cell as an osmotic system, i.e. membrane permeability, osmotic pressure and wall pressure, or by a metabolically-driven water uptake (active uptake).

Auxin does not appear to have any major effect upon the permeability of the cell to water; this has been demonstrated by measuring the penetration of radioactive water into the cell in the presence of the hormone. Neither is there any convincing evidence that the osmotic pressure of the cell sap prior to expansion is increased by auxin. On the contrary, in some tissues there is a fall in osmotic pressure preceding auxin-induced growth, and in others (e.g. wheat roots) an increase in osmotic pressure is associated with the inhibition of growth by auxin! Similarly, it has never been satisfactorily demonstrated that auxin promotes active uptake of water in growing tissues, although it is known, of course, that active metabolism is required for auxin-induced growth. It seems, therefore, that hypotheses attributing to auxin a direct effect on permeability, osmotic pressure or active uptake, cannot explain the promotion of cell expansion by the hormone.

Turning to the wall pressure, however, we find somewhat more favourable evidence that auxin might act here. For a decrease in wall pressure to occur, the wall must become more 'stretchable'. The 'stretchability', or deformation, of the wall has two components; one is the elasticity (reversible deformation), the other is the plasticity (irreversible deformation). The main approach, therefore, has been to measure these two components of the behaviour, under load, of untreated and auxin-treated tissues (e.g. hypocotyls or coleoptile sections). In experiments performed over relatively short times and before there was any growth, it has repeatedly been confirmed that auxin increases the plasticity of the tissue (figure 8.1). This increase in plasticity can be demonstrated even under conditions where subsequent growth cannot occur, for example in solutions of high osmotic pressure. Auxin, then, apparently 'softens' the cell walls making them more amenable to irreversible stretching – precisely the phenomenon which must occur in growth. The 'softening' of the wall leads, of course, to a drop in wall pressure, resulting in a lowering of turgor pressure; hence more water enters the cell and the wall is stretched. If we accept that this is the possible primary action of auxin in growth, how might it be achieved?

First, auxin may act directly on the wall and modify it in some way. Now, the structure of some auxins, e.g. IAA suggests a limited ability to bind metal ions. Such chemicals are called chelating agents and the phenomenon, chelation. How might this be relevant to the condition of the wall? Pectins constitute a considerable proportion of the wall, as the middle lamella and as part of the matrix in which the cellulose fibrils are embedded. The molecules of pectin may be envisaged as long chains having carboxyl radicals protruding at regular intervals. These are available for the formation of salt linkages, for example, by calcium forming a bridge between two carboxyl radicals of adjacent chains (figure 8.2). If this happens freely, a rigidity may be conferred upon the wall as the chains are now bound firmly to each other. Calcium is known to be present in cell walls (calcium pectate), and it is noteworthy that excess calcium, supplied from

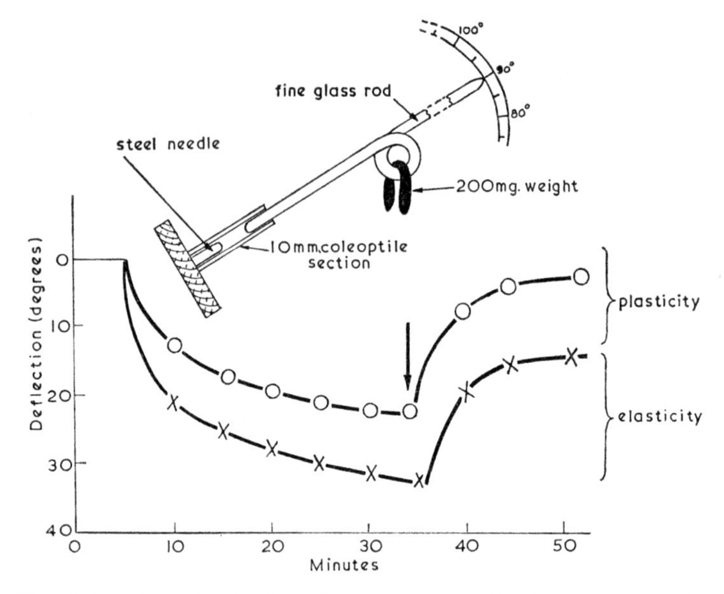

Fig. 8.1. The effect of auxin on the plastic deformation of a coleoptile. After treatment with an auxin solution (–x–) or water (–0–) the coleoptile section is arranged as shown and the deflection under load and after removal of the load (indicated by the arrow) is measured. The extent of recovery when the load is removed is a measure of the elasticity; the irreversible component of the bending is the plasticity. These two components are indicated for the auxin curve

outside, inhibits the growth of coleoptile sections and strongly reduces inherent and auxin-induced plasticity. One might imagine that auxin could enhance the softening of the wall by removing calcium ions, that is, by chelation. Unfortunately, there is no good evidence for this relatively simple view of auxin action; indeed, certain experiments using radioactive calcium indicate that there is no loss of the metal from the wall.

Some plant physiologists have speculated that auxin acts on the wall though the intermediary of enzymes, particularly those that methylate or demethylate the carboxyls of the pectin. Methylated carboxyl radicals clearly are unavailable for com-

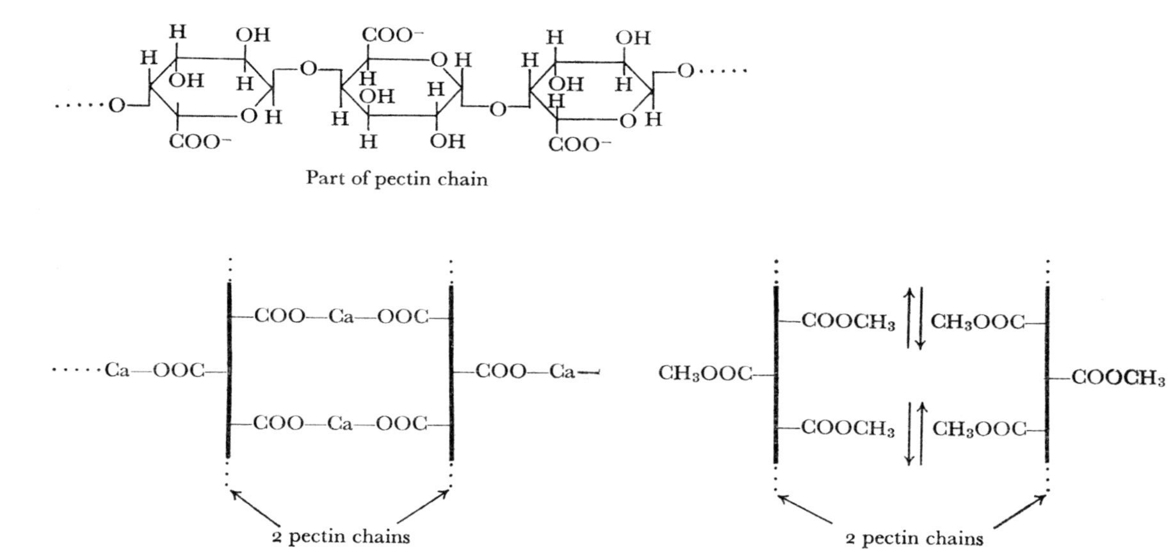

Fig. 8.2. The chemical structure of part of a pectin chain is shown. The carboxyls of adjacent chains may be united by calcium, and thus sliding of the chains is restricted; however, this is permitted when each carboxyl is methylated. (See note p. 58)

bination with calcium (figure 8.2). A small number of experiments with a rather restricted range of tissues has demonstrated that auxin possibly enhances methylation of certain pectin fractions of the wall; some lend support to the view that auxin might reduce demethylation, by acting on the enzyme which performs this process, pectin methyl esterase, but no clear picture has yet emerged.

This early phase of investigation on the auxin mechanism, preceding studies on protein and nucleic acid metabolism, thus provided strong evidence that auxin enhances wall plasticity but left the details of the mechanism rather obscure. We are now, however, at the exciting beginnings of renewed attempts at biochemical explanations. The idea is growing that auxin must affect enzymes involved in growth not so much by direct activation or inactivation but by inducing their synthesis. On this basis, the control exerted by auxins is at the heart of cellular activity – protein synthesis.

It has been known for some time that protein synthesis accompanies auxin-induced cell expansion but now it seems that synthesis of protein is actually necessary for this expansion. This conclusion derives from experiments using certain chemicals which specifically inhibit protein synthesis. Auxin has little or no growth-promoting action in the presence of these inhibitors. Further, in some tissues, it has been shown using radioactive tracers that auxin stimulates incorporation of amino acids into protein; protein synthesis, then, seems to be accelerated. It is not appropriate, here, to discuss the details of protein synthesis (see Chapter 3). We could enquire, however, as to the level at which auxin might act, since the synthesis of protein is the culmination of an elaborate series of events involving the synthesis of messenger RNA by the nucleus and the construction of polypeptide chains on the ribosomes using this RNA template. Although the evidence is by no means conclusive, certain experiments do suggest that auxins may act at nuclear level, causing the synthesis of new messenger RNA. Precisely how they would operate to achieve this is unknown.

Auxin, then apparently has a dual action: it increases wall plasticity and probably also the synthesis of protein. Can we connect these two effects with each other? It has now been found that inhibitors of protein synthesis (e.g. actidione) and of RNA synthesis (actinomycin D) almost completely prevent the action of auxin on plasticity. We might suggest, therefore, that the protein which auxin is suspected to induce is, or contains, the enzyme responsible for softening the cell wall. It is interesting that work with radioactive amino acids suggests that much of the auxin-induced protein is in the cell wall itself. Thus, if auxin does indeed induce the 'plasticising enzyme', the latter might well become bound into the wall. Here, it acts on a component of the wall and softens it, ready for expansion. As we saw earlier, the pectins have been popular candidates for the 'softening-up' process. Recent work does not particularly favour this view and the induced enzyme could even act directly on the cellulose or hemicelluloses of the cell wall.

As far as the growth-promoting action of gibberellin is concerned, the story is even less well documented. These hormones have been found to affect plasticity in a manner similar to that of auxin, but the mechanism is also unknown. We must remember that auxin seems to be necessary for gibberellin-induced cell expansion and so it is possible that any increases in plasticity observed after gibberellin treatment might really be due only to auxin, whose action the gibberellin is enhancing. Alternatively the explanation may lie with the known effect of gibberellins (in plants of a number of species) to increase the level of extractable auxin. Further information concerning the effect of gibberellin on cell growth is that continued synthesis of DNA might be necessary for the hormone to act. Why this is so is quite unclear. But attempting to explain the striking effects of the gibberellins on plant growth is not just a matter of understanding how these hormones cause cell elongation. Gibberellins also increase cell division (Chapter 6) and until the latter process is itself more perfectly understood we cannot begin to appreciate how gibberellins function.

In spite of these uncertainties there is considerable evidence that gibberellin, like auxin, influences protein synthesis. The evidence comes not from studies on growth, but on the action of gibberellic acid in germinating cereals, especially barley and wild oats. However, we may be justified in extrapolating from these findings in order to speculate on how the hormones could act in cell growth and division.

During germination of barley the starch in the endosperm is hydrolysed to sugars which pass into the growing embryo. The initial cleavage of the long starch molecule is performed by the enzyme, α-amylase. The action of this and other enzymes erodes the endosperm until several days after germination it may have completely disappeared. The 'digestion' of the endosperm visibly commences around the periphery of the grain and proceeds inwards. The α-amylase is, in fact, secreted by the layer of active, living cells around the starchy endosperm – the aleurone layer. However, if the embryo is cut off the grain, the aleurone fails to make significant amounts of α-amylase, and, indeed of many other enzymes. The 'influence' coming from the embryo which stimulates the enzymic activity of the aleurone is now known to be gibberellin.

One can see this remarkable action using pieces of endosperm surrounded by aleurone or even with isolated aleurone. When half grains of barley (i.e. with no embryo) are incubated in solutions of gibberellic acid for 48 hours, relatively massive amounts of reducing sugar are liberated, compared with those pieces in the control, i.e. in water alone (figure 8.3). This is due to the greatly increased amount of α-amylase coming from the aleurone layer. The sensitivity of the aleurone to gibberellic acid is phenomenal. A concentration of 10^{-10}M GA_3 is measurably effective, and when 1 ml of GA_3 solution is used, this, therefore, contains only approximately $3 \times 10^{-5}\,\mu g$ (i.e. 3×10^{-11} g) of the acid, of which, presumably, only a fraction actually enters the aleurone cells. Small pieces of aleurone may also be used, which are placed on sterile agar containing starch, plus, in the experimental dishes, gibberellic acid (e.g. $10^{-3}\,\mu g$/ml). After 2–3 days, iodine solution is poured over the gel, which becomes intensely blue except around those aleurone pieces supplied with GA_3; here, the starch has been broken down (Plate 10).

In similar experiments, the effect of GA_3 can be prevented by the inhibitors actinomycin D and actidione (see above). Moreover, if a radioactive amino acid is introduced together with the

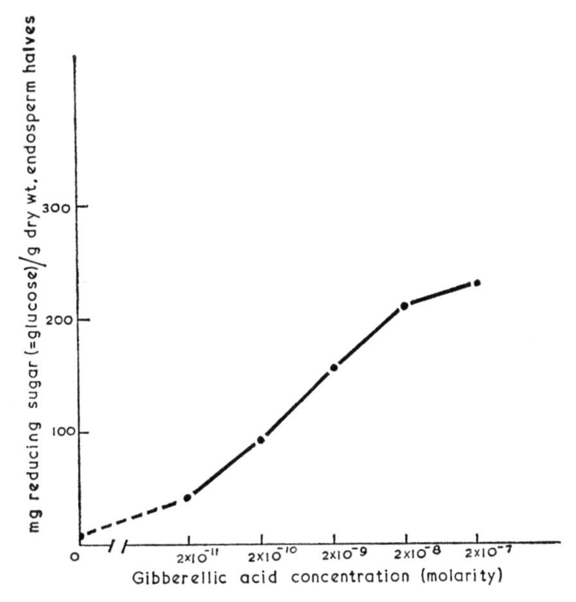

Fig. 8.3. The effect of gibberellic acid on the liberation of reducing sugar by barley endosperm halves. The GA apparently induces the formation of α-amylase in the aleurone layer. This enzyme is secreted into the endosperm and hydrolyses the starch to reducing sugar

GA_3, the α-amylase produced by the aleurone is found to contain the label. These two findings furnish evidence that the enzyme molecules are newly synthesised under the influence of the hormone. The ability of the gibberellins to promote protein synthesis in this system is, then, fairly well established. Future research will tell us if they act in this way in growth but one can conjecture that they may cause the synthesis of certain 'key' enzymes.

Other hormones – the cytokinins and the inhibitor, abscisic acid – are also implicated in nucleic acid and protein metabolism. The cytokinins, being largely purine derivatives, could plausibly feature in some aspect of nucleic acid metabolism; in fact, they are thought by some to be involved in the formation of a particular

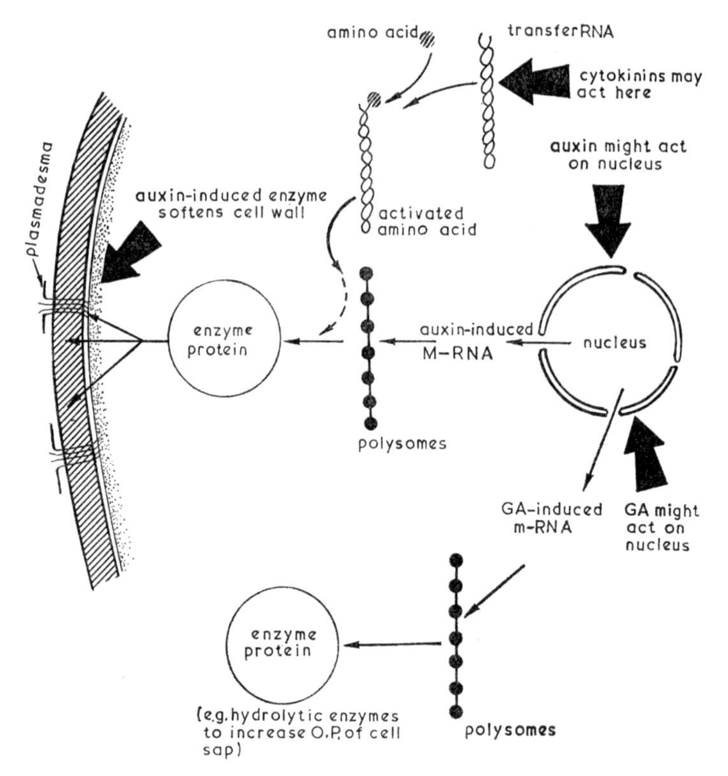

Fig. 8.4. A speculative scheme of the mechanism of action of the phytohormones in cell growth. The possible cellular 'targets' for the hormones are indicated

species of transfer RNA (see Chapter 3). Abscisic acid interferes with the action of GA_3 on barley aleurone and also inhibits enzyme synthesis in artichoke tuber tissue. Claims have been made that it stops DNA synthesis in *Lemna* fronds, and that this can be relieved by kinetin.

The auxins, gibberellins and cytokinins possess other properties which should be mentioned in this account of their possible mechanisms of action.

Auxins and cytokinins have an interesting action in causing

the mobilisation of various materials. For example, radioactive phosphorus is markedly attracted from the leaf to which it has been added, to the plant apex, or to the auxin-containing lanolin applied in place of the apex. The operation of this mobilisation phenomenon in growth control by auxin is not clear but its importance in apical dominance (Chapter 7) is obvious. Transport caused by kinetin was shown in an interesting experiment in which a radioactive amino acid was attracted, through the cells of a senescing leaf, to the position where kinetin was applied. This is thought to be of significance in connection with the ability of kinetin to retard leaf senescence. In certain species, gibberellins also delay senescence. Senescence, we should note, is characterised by a greater rate of protein breakdown than protein synthesis. Once more, then, the hormones are implicated in protein metabolism!

All these recent investigations have fascinating implications as they place the mechanism of phytohormone action at a fundamental level of cell biology – gene action and enzyme synthesis. In very broad terms we may conjecture that these hormones act as the keys for unlocking the genetic potential of the cell – to enable it to make certain enzymes, and subsequently to enlarge or divide. This would be particularly satisfying to one interested in comparative physiology or biochemistry for it points to the fundamental unity in the living world; many animal hormones, such as oestrogen and thyroxine operate in exactly the same process – in the control of protein synthesis. However, it is difficult to reconcile this hypothesis with all the known actions of the plant hormones. Auxin, for example, causes a large increase in protoplasmic streaming within just a few minutes after its application to certain plant cells. To interpret this in terms of 'gene activation' does not seem possible, at least at present.

We cannot yet give an unequivocal answer to the question 'How do hormones work?', but we see that some important concepts are emerging. We have to extend our investigations, not only as far as each individual phytohormone is concerned but also to the interactions among them, for as we have seen earlier, growth is likely to depend not on one hormone, but on several.

The solution of these problems is undoubtedly one of the major tasks for the developmental physiologist.

Summary

It is difficult to understand how hormones work particularly since they have so many effects on plant growth; nevertheless, current views implicate these substances in the control of nucleic acid and protein synthesis. In cell enlargement, for example, auxin may be involved in stimulating the production of an enzyme which makes the wall more stretchable. There is much evidence that gibberellins and cytokinins can also modify protein (i.e. enzyme) synthesis, but we are still far from understanding how this results in the promotion of cell growth or division.

Part 3: The external control

A variety of external factors strongly influence plant growth and development. Obviously, water supply and availability of mineral salts are of great significance; plants living under a water deficit or with a mineral deficiency make very poor growth. Light is of paramount importance since sufficient photosynthesis must occur to provide the sugars required for respiration, cell wall and protein synthesis and other metabolic reactions. These effects of the environment clearly must not be overlooked but as we have mentioned before, their actions are not confined to the growth process but rather they work indirectly, as it were, through the nutrition of the plant.

The environment does, however, exert specific control over plant growth. Sometimes the effect is so great that the onset of a particular phase of development (e.g. flowering, dormancy) is regulated by factors external to the plant such as temperature and daylength. In other cases, there is not an all-or-nothing response but rather the quantity of growth is determined. We can see this in the action of light on stems and leaves.

In Part 3 we shall be concerned with light, gravity and temperature as factors controlling growth. Although these three factors also control development, often by operating through a similar mechanism to that in growth, we shall leave the details of these effects until Part 4 of this book.

It is important to realise that external factors act through the internal mechanisms which control growth. In the same way, the driver of a motor-car exerts his influence only through the vehicle's inbuilt controls.

9

Growth and the environment

Light

Two plants – one grown in darkness and one kept in sunlight – present us with dramatic evidence for the control of growth by light. The former is spindly, with greatly elongated internodes, small leaves and a hooked apical bud (plumular hook) (Plate 11). It is also yellow (chlorotic) because chlorophyll synthesis has not occurred in darkness. This plant is said to be etiolated. With no chlorophyll, and in darkness, there obviously can be no photosynthesis, and once the plant exhausts its food reserves, death quickly follows. On the other hand, the light-grown plant is green, with sturdy, shorter internodes; the leaves have expanded and the apex is upright. As far as these plants are concerned there are, then, two effects of light, one connected with photosynthesis and one more specific to growth; the latter is manifest in the shorter internodes, larger leaves and straight apical buds. In this chapter we are concerned with these specific growth effects.

In these examples we see that the development of form (morphogenesis) is controlled by light; we use the word photomorphogenesis to describe such a process. This term embraces a number of other light-dependent events in plant growth and development which are listed in Table 2, but many of these (e.g. flowering) will be discussed in subsequent chapters.

The three photomorphogenetic responses described above are of considerable biological importance. One might argue that the failure of leaves to expand in darkness fits the shoot of a seedling well for its passage through the soil as they offer less resistance. Similarly, the unfolded apical bud is the best shape to protect the delicate meristem from the soil particles.

Table 2

Photomorphogenesis

Some of the many physiological processes in which light is involved

Phytochrome system	*Prolonged energy reaction*
Seed germination	Inhibition of germination
Flowering	Stem and hypocotyl growth
Leaf expansion	Cotyledon expansion
Stem growth	Anthocyanin synthesis
Hypocotyl growth	
Chloroplast orientation	
Photonasty	
Plumular hook opening	
Lateral root initiation	
Bud dormancy	
Tuber formation	
Sex expression	
Chlorophyll formation	
Anthocyanin synthesis	
Hair formation on hypocotyls	
Fern spore germination	

In the control of internode elongation, leaf expansion and straightening of the apical bud (the plumular hook), cell elongation is involved. Light apparently partially inhibits this in the internode but promotes it in the leaf and apical bud, differentially in the latter. We are immediately faced with the problem of how to understand why light has these apparently opposite actions. This problem is even more puzzling when we learn that all three cases are examples of the working of the same physiological system, called the phytochrome system. Let us now consider the characteristics of this system.

The phytochrome system

To effect a biological process light is first absorbed by a substance called a pigment. If the substance absorbs the red-green regions of the visible spectrum and reflects blue light, it is then a blue

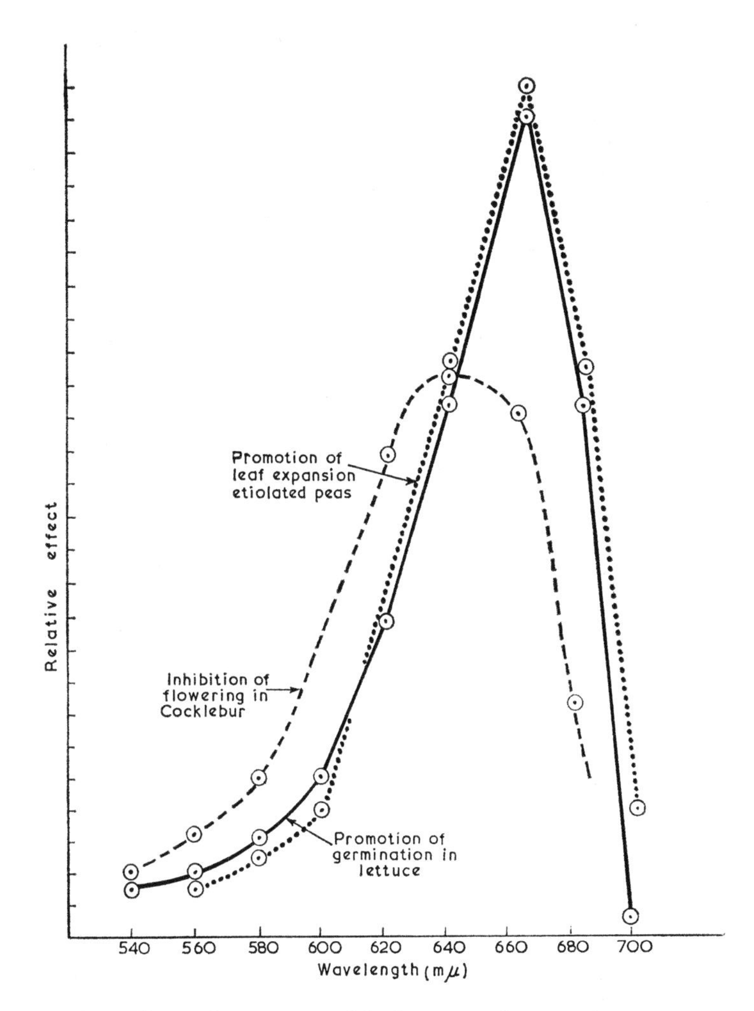

Fig. 9.1. The action spectra of 3 photomorphogenetic processes. Note the very close agreement and that peak activity is found in the red region of the spectrum (at 660 mμ in lettuce seeds and etiolated peas). The slight shift in cocklebur leaves is probably due to interference by the chlorophyll pigments, which are, of course, absent from lettuce seeds and etiolated peas

pigment. Clearly, we can learn something about the pigment(s) involved in photomorphogenesis by finding which wavelengths (i.e. colours) of light are effective, i.e. we determine the *action spectrum*. When we do this for various light-controlled responses, we discover a surprising fact. Processes as diverse as leaf enlargement, internode extension, flowering, seed germination and unfolding of the plumular hook are all controlled by almost exactly the same wavelength; that is, they have nearly identical action spectra (figure 9.1). Red light, with peak activity at 660 mμ wavelength, is effective, and so the same red-absorbing pigment is apparently involved in all these photo-responses. Very low light intensities are operative, of quite a different order from those required in photosynthesis, and often irradiation for only a few seconds or minutes is quite sufficient.

A variety of lettuce seeds (Grand Rapids) was used to find the action spectrum for germination. Most other varieties of lettuce will germinate in complete darkness; however it was found that these could be inhibited by light of the wavelength 730 mμ, i.e. far-red light. It seems, therefore, that red and far-red light have antagonistic actions. This finding led to a further fascinating and most important advance. It was discovered that when Grand Rapids lettuce seeds were exposed to a sequence of alternating red (660 mμ) and far-red (730 mμ) exposures, and then returned to darkness, germination subsequently occurred only when the final irradiation of the sequence was red light (Plate 11). This experiment shows that red light activates the pigment, which then can cause germination, but that the activity of the pigment is lost when it absorbs far-red light. Red light, in fact, converts the pigment into a far-red absorbing form. The action of the alternating irradiations shows that the two forms of the pigment are interconvertible by red and far-red. The same photoreversibility by red and far-red was shown for the other responses besides germination. For example, some hours after exposure to 1 minute of red light, leaves grown in darkness begin to expand; but if the red light is followed immediately by a few minutes far-red, no expansion occurs. This pigment system is called phytochrome and the two forms are designated as P_{660} (the red-absorbing form) and P_{730} (far-red absorbing).

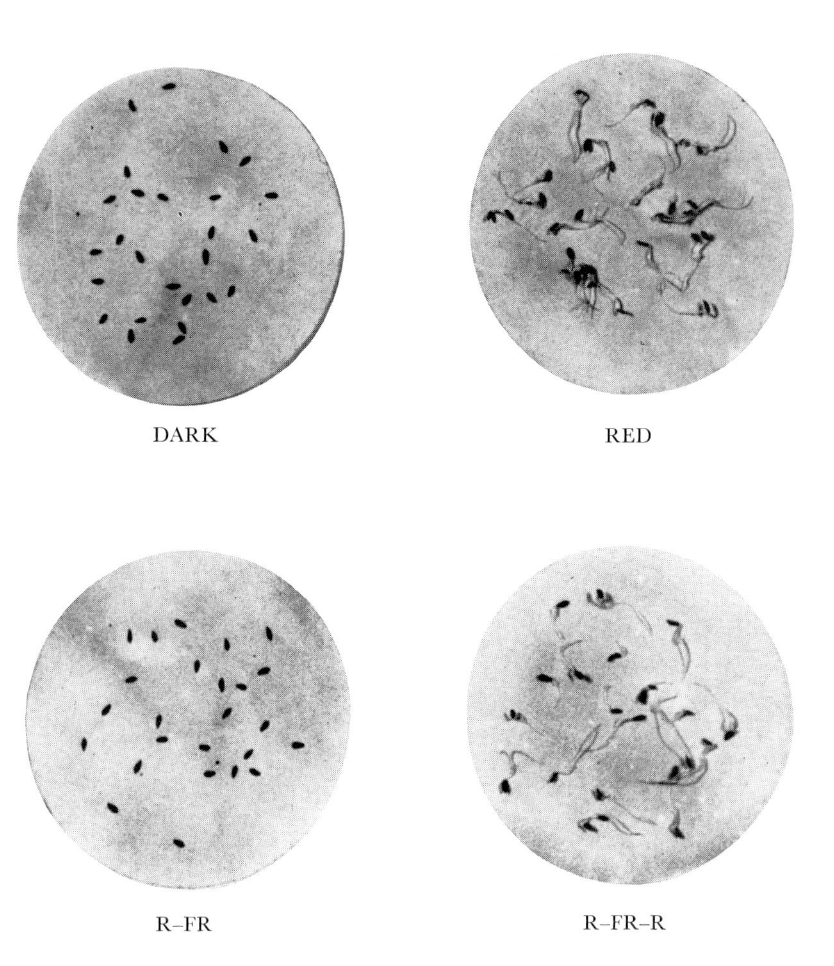

Plate 11 *The action of red and far-red light on germination of lettuce (var. Grand Rapids).*
Note that very few seeds germinate in darkness but almost complete germination results from a red-light treatment. Far-red nullifies the action of red, but it can be reversed by further red light. Germination, in fact, depends on the last treatment of the sequence.

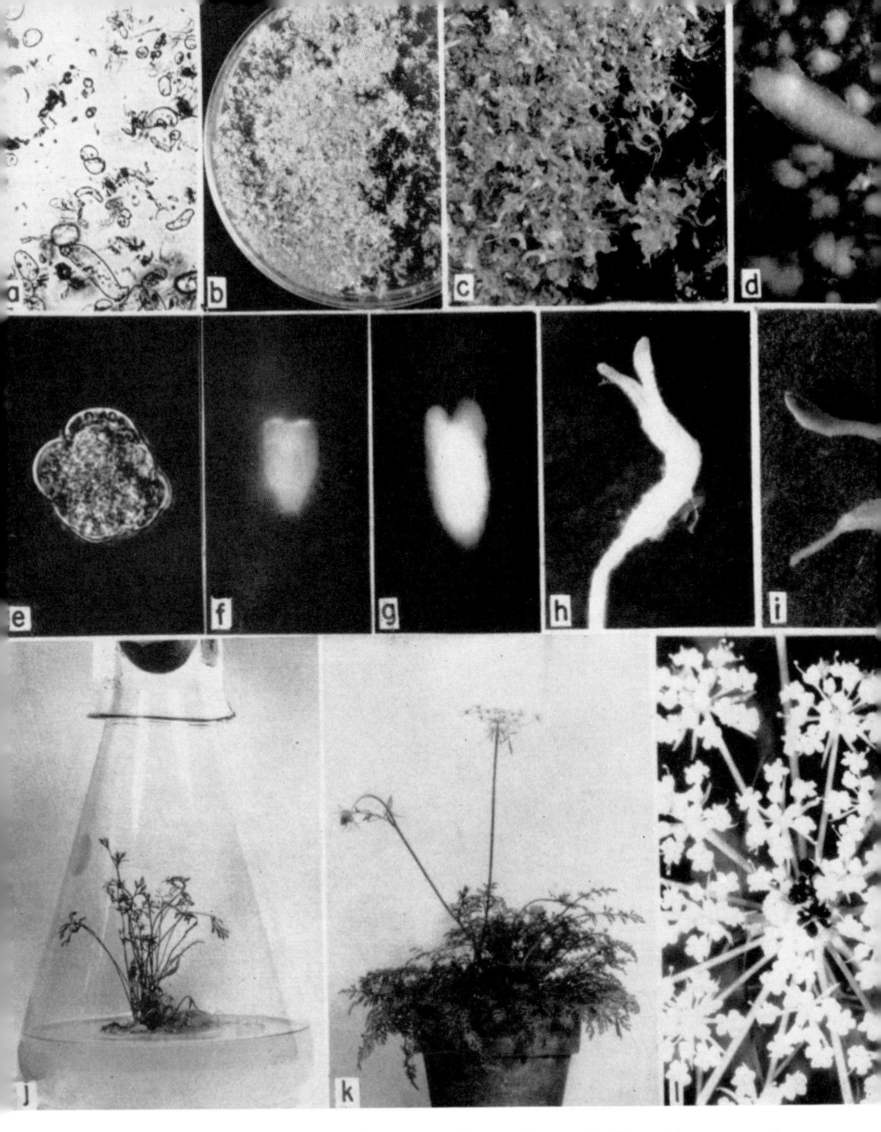

Plate 12 *Development of carrot plants from single cells grown in culture.*
(*a*) Free cells growing in liquid culture (*b*) and (*c*) Development of masses of embryos from free cells placed in a Petri dish. (*d*) Embryo development in a liquid medium. (*e*)–(*i*) Stages in the development of an embryo from a single cell (*e*). (*j*) and (*k*) Seedling and adult plant reared from single cells grown in culture. (*k*) Close-up of inflorescence of such plants.
The single cells were originally derived from a carrot embryo. Cells derived from carrot root phloem exhibit essentially the same phenomena.

Plate 13 *Shoot formation on calluses grown from pieces of tobacco pith.* The calluses on the right are growing on an agar medium containing a relatively high kinetin/auxin ratio and have formed shoots.

Plate 14 *The effect of photoperiodic conditions on flowering.*
Note that the tomato plants form flowers in long or short days although vegetative growth is greater in long days.

In a dark-grown plant or seed, then, phytochrome exists largely as P_{660}. The conversion $P_{660} \rightarrow P_{730}$ requires considerably less energy than the reverse so that even in light containing both red and far-red wavelengths (e.g. sunlight) the equilibrium is far to the right. This is why sunlight or artificial white light often have the same action as pure red light.

Further experiments whose details need not concern us here, suggest that in some species P_{730} can also revert very slowly (over several hours) to P_{660} even in darkness. This slow dark reversion is a most important property of phytochrome which we must appreciate to understand the role of light in growth.

Phytochrome, whose existence we see was postulated on the basis of purely physiological experiments, can now be measured directly in plant tissues by spectrophotometry. It occurs in minute amounts but it has, nevertheless, been extracted and collected in a test-tube. It appears blue and each molecule consists of two parts – a protein combined with a non-protein, light-absorbing part. The latter is a substance called a bilitriene, and

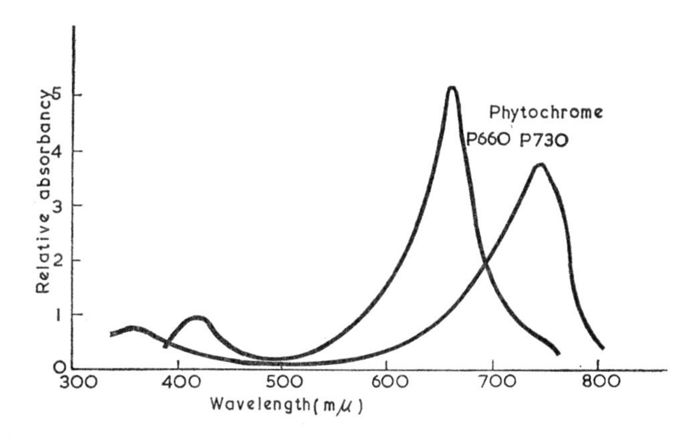

Fig. 9.2. The absorption spectra of the two forms of phytochrome. Maximum absorption by P_{660} is at 660mμ and by P_{730} at 730 mμ; but note that both forms absorb appreciably at about 700 mμ

is chemically similar to the bile pigments of animals and to the pigments of certain algae (phycobilins).

The technique of spectrophotometry has elucidated other properties of phytochrome which can be stated briefly as follows:

(*a*) Figure 9.2 shows the absorption of light by P_{660} and P_{730}; the curves overlap in the far-red. This means that even in far-red light some P_{660} is transformed to P_{730}.

(*b*) P_{730} is a very labile form of the pigment. Not only does some slowly revert to P_{660} in darkness but considerable amounts are destroyed (figure 9.3).

(*c*) New P_{660} can be synthesised; there is evidence that this synthesis may itself be light-promoted.

We can now summarise the properties of the phytochrome system;

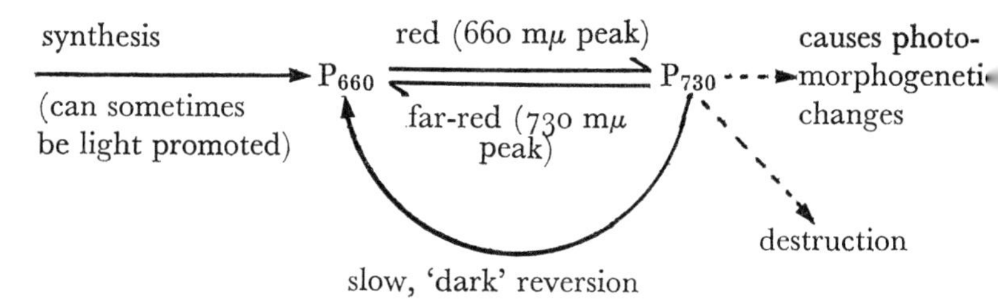

We have already mentioned that the light-induced conversions of phytochrome occur quite rapidly. It is not necessary, therefore, to illuminate plants continuously to obtain some degree of response. Just 2–3 minutes of light cause the plumular hook of a dark-grown seedling (e.g. French bean) to open during the next day or so, and the leaves to expand perhaps 100% over the dark control. Similar short exposures suppress mesocotyl growth in certain cereal seedlings and inhibit growth of internodes, epicotyls and hypocotyls of numerous plants. In many cases, of course, illumination for a longer duration or at daily intervals is more effective. This is connected partially with the necessity for P_{730} to be present for rather a long time or to be replenished daily. Neither of these will happen after a single light 'flash' of 2 minutes since dark reversion and destruction of P_{730} both occur.

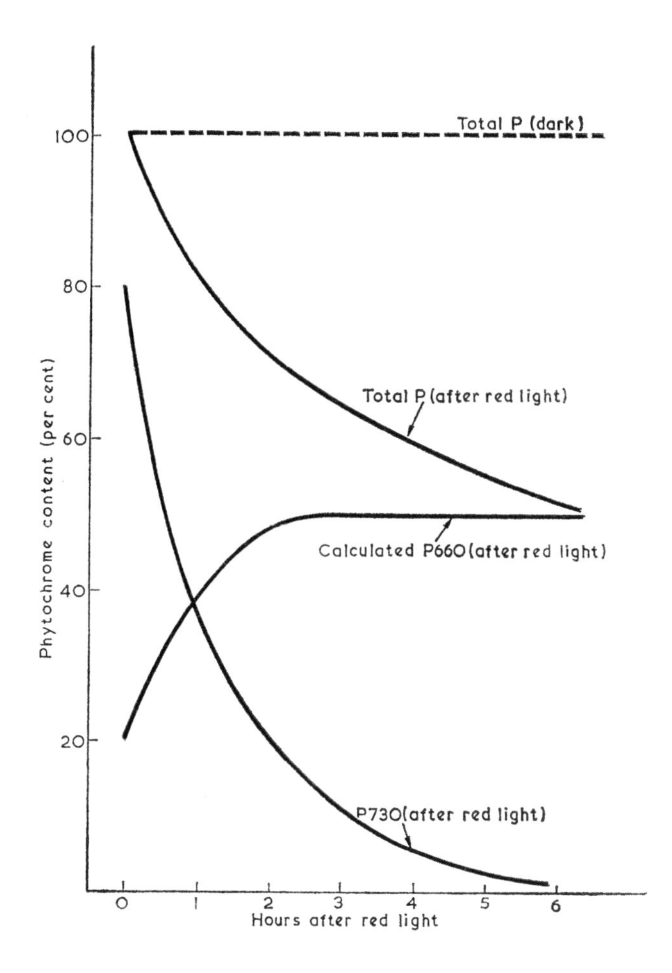

Fig. 9.3. The change in phytochrome content after irradiation of etiolated pea stems with a few minutes red light. There is a rapid drop in the level of P_{730}. This is due partly to reversion to P_{660} (which shows an accompanying increase) and partly to destruction of P_{730} in darkness. Thus, the total level (total P) also drops. (after Furuya and Hillman)

The prolonged energy reaction

The phytochrome system is not the only one involved in photomorphogenesis. We now know of a multitude of responses which are governed by irradiation for relatively long periods (i.e. several hours) especially with far-red, but also with blue light. The system involved is called the prolonged (or high) energy reaction (figure 9.4 and Table 2).

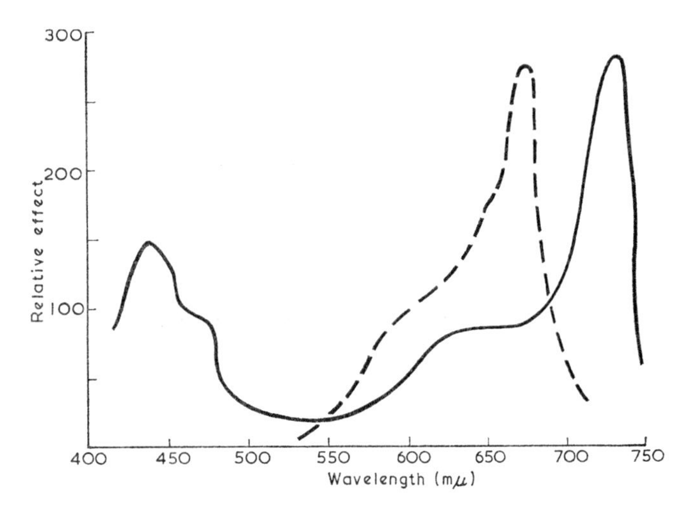

Fig. 9.4. The action spectrum of the prolonged (high) energy reaction. The continuous line is the action spectrum for this reaction in the control of cotyledon expansion in white mustard; the dashed line is the action spectrum for a phytochrome-controlled response. (after Mohr)

We can distinguish three categories of the prolonged energy reaction: (*a*) Prolonged far-red or blue light can have the same effect as a short exposure to red light. For example, cotyledon expansion in white mustard, and suppression of internode elongation in some species, are effected by either a few minutes of red light or several hours of far-red or blue. (*b*) The second type of response is where prolonged irradiation has the opposite action to a short duration of red light. The lettuce seedling is an example of this. Dark-grown seedlings have no plumular hook but it can be induced by red light. However, once it has formed it can then be reopened by exposure to blue or far-red light for a few hours. (*c*) The third category is when the prolonged energy reaction has exclusive control over growth. We can see this again in the lettuce seedling. In this species hypocotyl growth is unaffected by red light but is inhibited by prolonged blue or far-red illumination. As far as this seedling is concerned then, the inhibitory action of sunlight on hypocotyl growth is because of the far-red and blue components and not the red wavelengths.

It is rather difficult at present to understand the basis of these various aspects of the prolonged energy reaction. Some plant physiologists believe there are pigment systems sensitive to either prolonged blue or far-red or to both. Another possibility is that the prolonged energy reaction actually operates through the phytochrome system and that some of the growth responses (especially in (*a*)) are really caused by the continual presence of low levels of P_{730}. This is brought about because of the slight absorbance of far-red light by P_{660} (figure 9.2). Thus, a small amount of P_{730} is formed and even if this is destroyed more replaces it to restore the equilibrium. In prolonged far-red light, then, some P_{730} is always present.

Growth of stems and coleoptiles in light and darkness

We have seen that light partially inhibits elongation of stems and coleoptiles. The patterns of response of these organs has been investigated in detail in only a few species, but these reveal a surprising fact: that at first light really accelerates growth but this stops much earlier than it does in total darkness. Figure 9.5 shows this in the oat coleoptile. If a young dark-grown pea seedling is exposed to light, growth of the youngest internode is first promoted while the older internodes are inhibited! It seems, therefore, that light (i.e. phytochrome and prolonged energy reaction) stops cell division in the young internode, and cell expansion then begins. Cells which are already elongating, such as in the older internode, are accelerated but are induced to mature earlier. The combined effect of these two responses is an

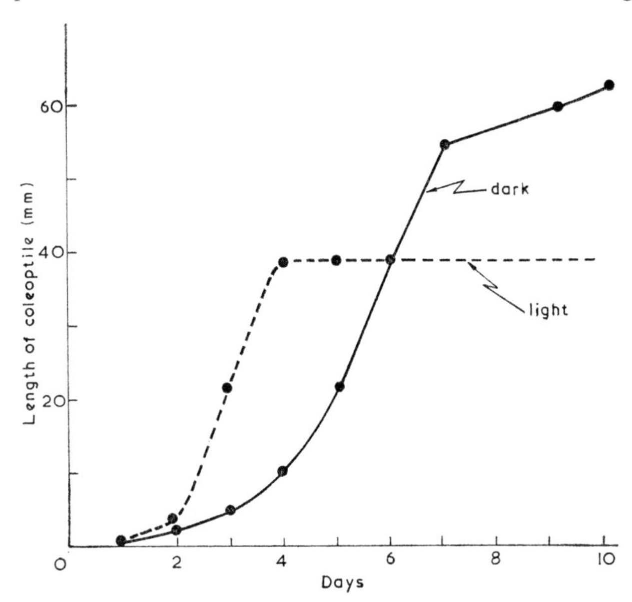

Fig. 9.5. The time course of growth of oat coleoptiles in light and darkness. (after Thomson)

inhibition of growth in the seedling as a whole. Clearly, the growth pattern is therefore fundamentally the same in light and darkness, but in the former growth occurs more rapidly and to a more limited extent.

Mechanism of action of light

How do the pigments involved in photomorphogenesis change the quality and quantity of growth of a stem, coleoptile, leaf or plumular hook? We cannot yet give a complete answer to this intriguing question but active research in the last few years has produced much information.

The photochemical reactions in photomorphogenesis, e.g. the conversion $P_{660} \rightarrow P_{730}$, are almost instantaneous, yet the morphogenetic results of this are discernible only after hours, days or even weeks. The ultimate response in growth, that of cell enlargement, presumably occurs in the manner we have discussed in

Chapters 3 and 8 and we need not be concerned here with further details. We must rather turn our attention to the sequence of events set in motion by light absorption and which culminate later in the observed response.

First, we should ask, 'What are the primary biochemical actions initiated by phytochrome and the prolonged energy reaction?' Now, although the observed growth responses to light might take several hours or days to occur, there is good evidence that P_{730} acts quite rapidly. Some other responses controlled by P_{730}, such as leaf and leaflet movement in *Mimosa* and chloroplast orientation in other species, are actually visible within minutes of the beginning of illumination. Even in systems where the morphogenetic response occurs hours after exposure to light there is evidence that P_{730} must begin to act quickly.

It is logical to assume that P_{730} has a common primary action in all the different systems in which it is involved. Because of its rapid action in *Mimosa* pulvini, where quick changes in cell turgor occur, it has been proposed that phytochrome affects permeability (in this case, to water) of cellular membranes. Since it is a protein it might even itself be part of a cell membrane. Absorption of light would change the shape of the protein molecule (just as such changes occur in the pigments of the retina of the eye) and thus the structure of the membrane might be altered. Of course, once the permeability of cell membranes is altered a host of biochemical reactions might then occur, for certain substrates will then be accessible for enzyme action. Such events might eventually lead to enhanced growth.

Another view is that both phytochrome and the prolonged energy reaction are involved in gene regulation. There is, indeed, some evidence of an effect of these photomorphogenetic systems on nucleic acid and protein synthesis. It would be attractive to postulate that light is concerned with the activation of specific genes in the cell nucleus, but such a hypothesis is still only tentative. Finally, since phytochrome is a protein it might itself be an enzyme catalysing a 'key' reaction in cellular metabolism. Unfortunately, we have no experimental confirmation for this idea, which is nevertheless quite possible.

Light and hormones

The primary action of the pigment systems is, then, still unclear, but we nevertheless know of some biochemical changes which occur between the absorption of light and the final growth response. In coleoptiles and pea epicotyls red light lowers the auxin content, evidently by enhancing IAA oxidase activity. In young internodes of some plants red light apparently leads to the formation of an inhibitor of IAA oxidase, whereas in older internodes a co-factor for IAA oxidase is formed. These changes fit in well with the responses of young and older internodes to light described on page 123.

Light apparently can lower the sensitivity of stems to endogenous gibberellin (page 83). How these hormones are involved in the other growth effects of light (leaf expansion, etc.) still remains to be resolved. Since leaf expansion in some species is enhanced in darkness by gibberellin treatment, it has been suggested that light causes gibberellin biosynthesis in leaves; indeed, there is now some evidence for this view. We might note, in this context, that light-gibberellin relationships present something of a paradox. In leaves, light and gibberellin have the same action – to promote growth; in stems, however, light inhibits growth and gibberellins promote it!

We are learning more of the ways in which light is involved in growth, but we do not understand the mechanism. Hormonal systems must almost certainly be involved but the detailed biochemistry of the light-induced changes is unfortunately still relatively obscure.

Phototropism

Another common effect of light, but one not normally considered under the heading of photomorphogenesis, is phototropism. This is where the direction of growth is determined by light coming from one side. Coleoptiles, stems and hypocotyls grow towards the incident light, i.e. they are positively phototropic. Roots of relatively few species grow away from unidirectional light, and are, therefore, negatively phototropic. On the other hand, leaves align themselves at a certain angle to the light.

Darwin's observations on grass coleoptiles made it clear that bending is due to differential growth of the cells several millimetres below the apex but that the apex itself perceives the light stimulus. The same has been shown to be true for stems. The bending response is clearly observed a few hours after exposure, provided a certain minimum total light energy has been received.

Determinations of the action spectrum for phototropism show that only blue light is effective (figure 9.6) and so a different

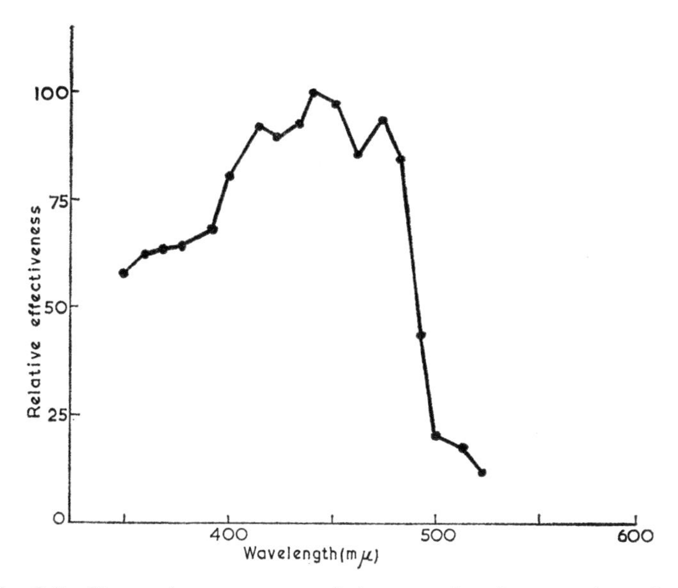

Fig. 9.6. The action spectrum of phototropism in oat coleoptiles (after Leopold)

pigment system from those discussed above is involved. The pigments must obviously be present in the apical region of the shoot or coleoptile. Two pigments have been suggested, which both absorb light in the blue wavebands. These are β–carotene and riboflavin, and there is controversy as to which is the pigment most likely to be involved. The fact that certain carotene-less mutants can show phototropic responses throws doubt on carotene

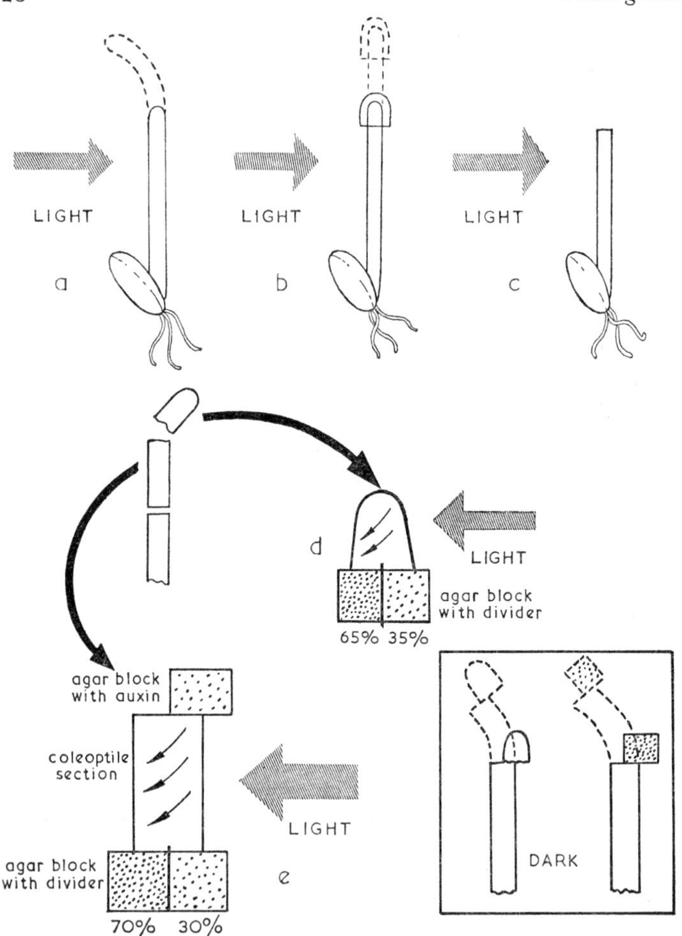

Fig. 9.7. Phototropism and auxins. Coleoptiles (and stems) grow towards the light but not if the tip is covered or cut off (*a*, *b* and *c*). Light seems to cause a redistribution of auxin to the darker side of a coleoptile apex (*d*) or of a coleoptile section supplied with auxin (*e*). The redistributed auxin diffuses into the collecting blocks of agar. In darkness, an eccentrically-placed tip or block of agar with auxin in it causes the coleoptile to bend as though it had been illuminated. This, too, is due to unequal distribution of auxin down just one side of the coleoptile.

as an acceptable candidate. However, β–carotene may be important (when present) in producing a gradient of light intensity within the tissues by acting as a kind of 'pigment screen'.

The phototropic response occurs in the growing cells behind the perceptive region. In shoots and coleoptiles greater growth occurs on the darker side, hence the bending. Auxin is clearly involved, as shown by experiments summarised diagramatically in figure 9.7. Moreover, tropic responses can be made to occur in auxin-deprived material only by the supply of auxin.

How is auxin implicated in this differential growth? Two theories have been advanced: light redistributes auxin to the darker side or light causes auxin destruction on the lighted side. In both cases, because of the polar movement of IAA, cells on the darker side came to acquire relatively more auxin. Auxin redistribution was first found by the early students of these hormones who collected more auxin from the shaded side of coleoptile tips (figure 9.7). This has been confirmed in recent times using similar techniques and also radioactive IAA. It now seems clear that radioactively-labelled IAA does migrate laterally under the influence of light.

The idea of auxin destruction is based partially on the experimental finding that riboflavin, on absorbing blue light, causes the destruction of IAA in tissues possibly by reducing the action of an inhibitor of IAA oxidase.

The phenomenon of phototropism is, then, still problematical. There is good evidence for the concept of auxin redistribution. Although we do not yet understand how this is brought about, light-induced bioelectric potential gradients are possibly very important; these may well be responsible for the migration of the charged molecule of IAA.

Photonasty

Nastic movements are due to reversible changes in cell turgor. They are not directional responses to stimuli nor are they growth movements and thus they differ on two counts from tropisms. Photonasty is seen, for example, in the movements of leaves and petals of some flowers. The light-induced changes in cell turgor which are involved, sometimes occur in special tissues (pulvini) which cause these movements. Some are phytochrome-controlled and are discussed briefly on page 125.

Gravity

The effect of gravity on plants is seen in two phenomena – gravimorphism and geotropism.

Gravimorphism is the term sometimes used to describe those effects of gravity on total growth and on development, particularly flowering. When certain shoots, for example larch, cherry and apple, are held horizontally, both the number and length of internodes are reduced, compared with the vertical controls. These horizontal branches also produce many more flowers and hence more fruit and seeds. This effect has long been put to good use and is the basis for downward training of branches of fruit trees. In gravimorphism, there may be a redistribution of auxin to the lower side, but it is doubtful if this fully explains the phenomenon.

When shoots and roots are displaced from their normal growing position differential growth occurs which causes the shoot tip to curve upward, the root tip downward, and the vertical positions are re-established. Many leaves behave similarly. This is geotropism. These responses follow when the organ is subjected to the gravitational stimulus for a certain minimum total time, the presentation time.

In general, roots are positively geotropic, growing towards the gravitational pull, but coleoptiles, and many leaves and shoots are negatively geotropic. Shoots clearly exhibit various grades of response. Secondary, tertiary and subsequent branches do not grow vertically and neither do underground or creeping shoots (e.g. rhizomes and runners).

The change in orientation is brought about by differential cell elongation (figure 9.8). Cell growth is greater on the uppermost side of a horizontally placed root, but on the lower side of a stem or coleoptile. It is likely that auxins are involved and indeed in coleoptiles and hypocotyls the hormone apparently accumulates on the lower side (figure 9.9). In some cases there is also evidence that auxin synthesis is modified by change in position of an organ, or that sensitivity to auxin might be affected. Further, for reasons which are completely unknown, both red light and gibberellins can affect the geotropic response.

The perception of gravity in geotropism occurs in the very tip of shoots and coleoptiles, and in roots, probably in the root cap itself. Most theories propose the displacement by gravity of relatively large intracellular bodies, composed of starch, called

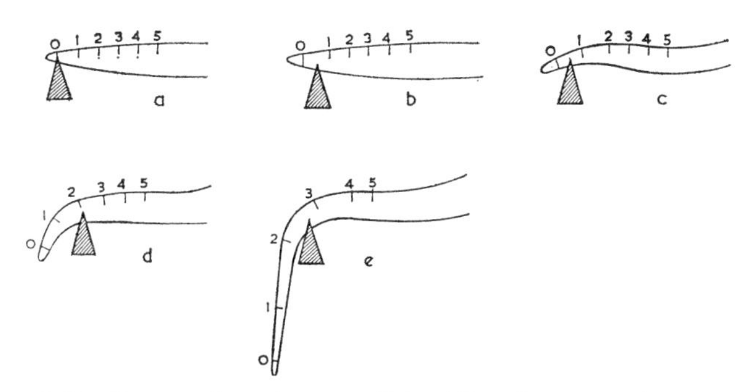

Fig. 9.8. Geotropism in a radicle. The downward curvature occurs because of greater growth on the upper side of the elongating zone 1–4. (after Sachs)

statoliths (figure 9.10). In a manner which is not yet understood, this displacement then causes the various changes of auxin level and hence a curvature.

Temperature

This controls growth by affecting the rates of all metabolic reactions. These are increased at warmer temperatures and overall growth is therefore accelerated. However, this frequently produces a spindly and weak plant which shows poor flower development. Evidently, lower temperatures are often necessary to produce 'normal' growth. For example, the best conditions for the tomato are night temperatures of about 10 °C and day temperatures around 20 °C. The effect of daily temperature fluctuations on general growth and flowering is sometimes called thermoperiodism.

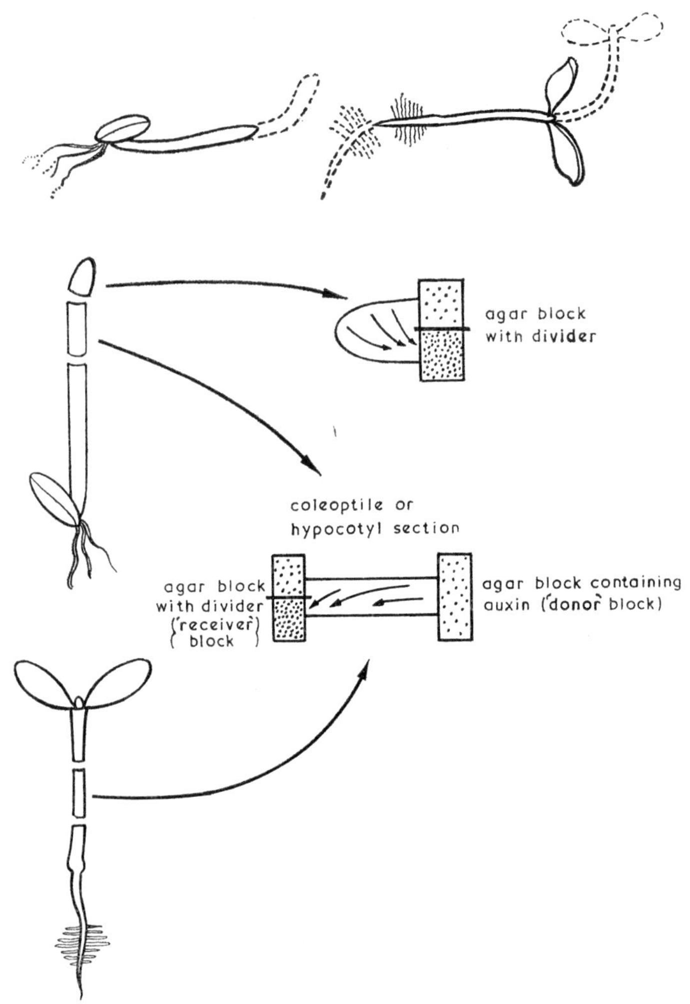

Fig. 9.9. Geotropism and auxins. Coleoptiles and stems show negative geotropism i.e. they turn upwards from a horizontal position, whereas roots show positive geotropism. Gravity apparently causes redistribution of auxin in a coleoptile apex and in coleoptile or hypocotyl sections. The redistributed auxin can be collected in agar blocks

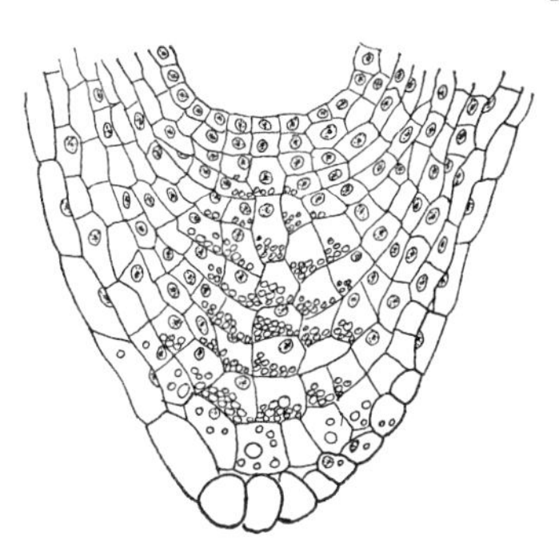

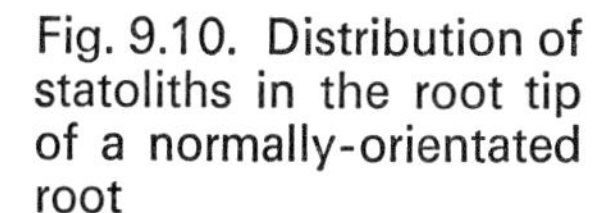
Fig. 9.10. Distribution of statoliths in the root tip of a normally-orientated root

In addition, an important control of plant development is exerted by temperatures a few degrees above freezing. Flowering and dormancy, for example, are profoundly influenced by such conditions; these are best discussed in the relevant chapters in Part 4.

Summary

The phytochrome system (red and far-red light) and the prolonged energy reaction (far-red and blue) control photomorphogenesis being involved in such diverse phenomena as leaf expansion, stem elongation, seed germination, plumular hook opening and photoperiodism. Another pigment system which absorbs blue light controls phototropism. The mechanism of action of these systems is unknown but in some cases they may influence internal hormone levels, by effecting their synthesis or destruction, or their distribution (e.g. auxins in phototropism). Auxin redistribution is probably also caused by gravity; this accounts for geotropism and gravimorphism.

The temperature influences growth by affecting metabolic rates, but 'normal' growth often requires daily temperature fluctuations. A most important developmental control is achieved by temperatures just above freezing.

We must conclude that growth depends on an interplay between external and internal factors, in a highly ordered and organised system.

Part 4: The control of development

So far we have looked mainly at quantitative changes in growth, i.e. increase in size. However, even during increase in size the constituent cells are changing in type, i.e. they are differentiating into xylem, phloem and all the other kinds of cells which comprise the particular organ. Such cellular changes are not immediately evident morphologically. But differentiation does occur to produce obvious changes, i.e. new organs such as leaves, flowers, fruits, etc. In the following chapters we deal with this subject of development.

10

The control of differentiation

Morphogenesis

Even before we begin to examine plants scientifically we are aware of certain basic features of their structure. We know that they consist of stems and roots and that the former bear the leaves, buds and flowers. Moreover, we can recognise particular species by their characteristic leaf shape and arrangement upon the stem. Later, when we study plant anatomy, we learn of the many different types of cells and how they are arranged in the plant body. But why *does* only the stem apex, not the root apex, form leaves and buds with a distinct arrangement; and what determines whether a cell shall differentiate into phloem or xylem? These fundamental problems concerning the development of form (morphogenesis), and differentiation, are perhaps the most challenging kinds in modern biology. Although they still remain largely unanswered, we are beginning to discover more about the control mechanisms involved.

Differentiation begins with the very first division of the fertilised egg. This is clearly seen in the alga, *Fucus*. Here, the newly-formed zygote develops a protuberance on one side and this is part of the smaller cell which is then cut off (figure 10.1). This cell *always* gives rise to the rhizoids, while the larger cell forms the main body of the plant. Thus, the destiny of the cell is determined just before division when the zygote acquires a *polarity* due to the swelling. A number of factors apparently determine this polarity, such as light, and pH, temperature and oxygen gradients.

Polarisation is thought to occur in the zygotes of higher plants which are, of course, retained within the ovule. The factors involved are unknown but we might guess that similar gradients may be present as well as nutritional and hormonal effects, all imposed by the surrounding tissues. When a fertilised egg of a flowering plant first divides the two daughter cells have distinct and determined destinies – one forms the suspensor and the other the embryo (figure 10.2). Early in the development of the embryo further 'determination' occurs when cells become either the incipient radicle or the cotyledons and apical meristem. Very early on then, even when there is no clear structural differentiation, physiological and biochemical changes have occurred which

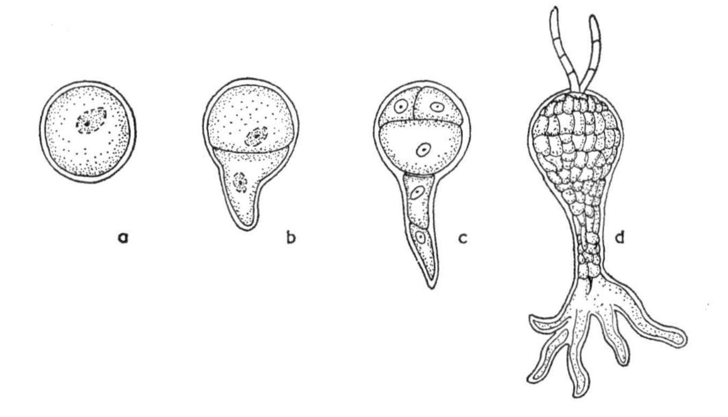

Fig. 10.1. The development of the embryo of *Fucus.* The cell with the protuberance (*b*) develops eventually into the rhizoids (*c* and *d*) (*d* is drawn to a smaller scale)

decide that a cell shall give rise to a root or a shoot. The changes are manifold and concern such features as whether or not the tissues shall subsequently be positively or negatively geotropic, shall be able to produce leaves, or still have the capacity to make certain chemicals.

It is axiomatic that all cells derived from the zygote will have the same chromosome and genetic complement. How does it happen, therefore, that some of the daughter cells have only root characteristics? One might speculate that although the cells start by being genetically identical, they later undergo a permanent genetic change during the course of their development. Good evidence against this view comes from many sources but one example, involving tissue culture, is particularly convincing. When small discs are cut from the phloem of carrot roots and placed in a nutritive medium containing mineral salts, vitamins, carbohydrate, auxins and cytokinins, the cells divide to form a mass of parenchymatous tissue called a callus. Cells which break

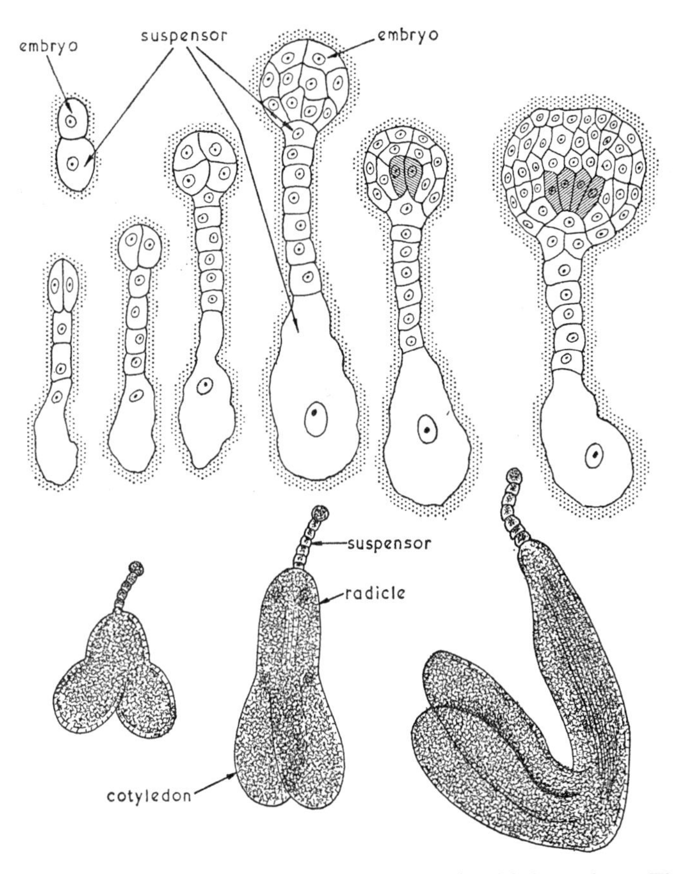

Fig. 10.2. Development of the embryo of a higher plant. The first division produces a cell which gives the suspensor and one which becomes the embryo proper. At quite an early stage in the latter's development cells are formed which are destined to become the root and shoot apices (the hatched cells)

off this callus, when transferred to another medium and under carefully controlled conditions, divide and begin to form a body very much like a young embryo (Plate 12). This continues to develop until eventually a perfect seedling forms which can be planted out and subsequently forms flowers and seeds. Clearly, then, all the information pertaining to stem, leaf and flower characters is still available, even in cells which are derived from a root.

The inference we can draw from this and similar experiments is that during differentiation and morphogenesis the capacities of cells become modified but not permanently changed. What factors are responsible for these modifications? The control mechanisms are likely to be quite complex – in fact the total cellular environment must be important. However, we do know that plant hormones play an extremely important role. This is demonstrated dramatically again by experiments using the techniques of tissue culture. We have already seen (Chapter 4) that tobacco pith cells divide and produce a parenchymatous mass of cells – a callus. Provided auxin and a cytokinin (e.g. kinetin) are supplied this continues for long periods. But if the relative concentrations of auxin and kinetin are varied, differentiation in the callus can be induced. At a relatively high auxin–kinetin ratio root primordia are formed which eventually grow out into mature roots. On the other hand, buds are induced by relatively high kinetin–auxin ratios, and leafy shoots grow out (Plate 13). These effects illustrate the control by hormones over organ formation, and also point once more to the ability of cells to 'dedifferentiate'. How hormones act in the control of organ formation in the intact plant is unclear. They are certainly involved in flowering (Chapter 11) and root initiation (page 94) but their role in controlling the pattern of leaf and bud development at the shoot apex remains to be elucidated.

The inception of leaf primordia at the apex occurs in a particular arrangement which is characteristic of the species (i.e. in spirals, whorls or opposite). What determines this pattern? In fact, why should leaf primordia be formed at all? There is, unfortunately, no ready answer to these problems. We might think in terms of hormonal control, in which areas of cells are induced to divide more rapidly and thus become pushed above the apex to form primordia; but why should this happen only at certain positions? Fields of 'influence' of already existing primordia may be important, and a new primordium can form only in the next 'free' position. We do not know what biochemical form these 'influences' may take, and we still await a satisfactory explanation for this aspect of morphogenesis.

Once a leaf has been formed what determines its mature shape? Obviously, the genetic factor is most important, but the environment can have profound effects. For example, in some aquatic plants the submerged leaves are quite different in shape from the aerial ones. Other influential environmental factors are light intensity, daylength and temperature. Presumably, these environmental factors work on the internal controls which may be auxins, gibberellins and cytokinins. These hormones might interact to cause cells of certain parts of the leaf to divide and expand more than others, and so the characteristic leaf shape is produced.

Cellular differentiation

We know relatively little about the control of cell differentiation in higher plants, i.e. why a cell becomes parenchyma, collenchyma, etc. However, in the case of vascular tissue the hormones are clearly involved. Division and differentiation of secondary cambium depends on auxin and gibberellin coming from near-by buds and leaves. Similarly, regeneration of xylem in wounds is induced by auxin from the leaves. Differentiation of parenchymatous callus cells into vascular tissue is promoted by a bud which is grafted on to it (figure 10.3), and this seems to be due to auxin. In fact, xylem cells are formed in callus tissue when a wedge of agar containing auxin is implanted. However, auxin alone does not control the position and extent of differentiation for the gradient of sucrose also exerts a strong influence. In the presence of 4% sucrose, a relatively high concentration, differentiation in callus occurs almost exclusively into phloem cells; with lower sucrose levels, xylem is formed. This effect of sucrose might explain the pattern of differentiation from secondary cambium.

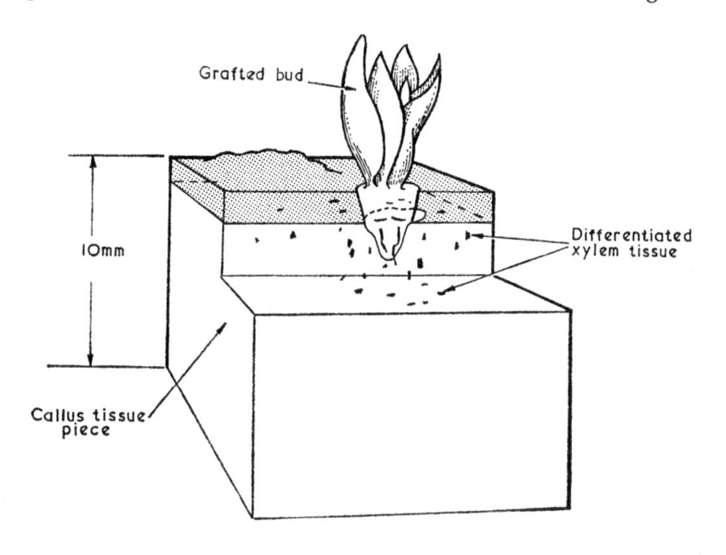

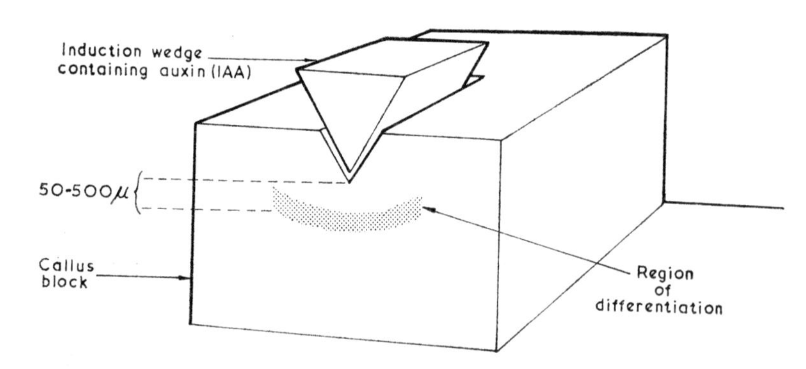

Fig. 10.3. Induction of xylem in callus tissue. The upper figure shows the differentiation of xylem elements induced in a callus by a shoot apex of *Syringa* grafted on to it. The lower figure illustrates a similar induction of differentiation induced in a callus by a wedge of agar containing auxin (after Wetmore and Sorokin, and Jeffs and Northcote)

Nucleus and cytoplasm in differentiation

So far we have dealt with factors outside the cell or tissue which determine differentiation. In the final analysis, though, no cell can differentiate if it does not have the capacity to do so, whatever the external influence is. We might expect that the nucleus plays the decisive role: what is the evidence for this?

As far as higher plants are concerned, there is no direct experimental work which bears upon this, other than the genetic investigations. This is because of the great technical and manipulative difficulties involved. On the other hand, fairly large, single cells of animals can easily be obtained (egg cells, protozoa) and here the role of the nucleus has been extensively investigated. Nevertheless, one plant has contributed much to our understanding of the control exerted by the nucleus in differentiation; this is the single-celled alga, *Acetabularia*. In the adult form (figure 10.4) the cell consists of a stalk (about 5 cm long) surmounted by a 'cap', and the major part of differentiation of the cell is concerned with cap formation. The nucleus is in the rhizoids at the base of the cell which can easily be cut off without damaging the remainder of the cell. Provided they are kept in the light, the stalk, without nuclei, so produced, grow and eventually form a cap, although such activity is limited. It seems that a 'morphogenetic substance' had already passed out of the nucleus into the stalk and this can support differentiation for some weeks. This morphogenetic substance is certainly RNA, or a precursor of RNA, which can later provoke differentiation only in the light. Grafting experiments with *Acetabularia* show more clearly the determinative role of the nucleus. Two species, *A. mediterranea* and *A. crenulata* have rather different cap shapes. When a rhizoid containing a nucleus of *A. mediterranea* is grafted on to the anucleate stalk of *A. crenulata*, the cap that is induced is intermediate in shape between the two types. When this cap is cut off, the new one that forms is entirely of the *A. mediterranea* type. Thus, immediately after the first graft 'morphogenetic substances' of both species are present which both exert their effect, but after the first cap formation only that of *A. mediterranea* is present

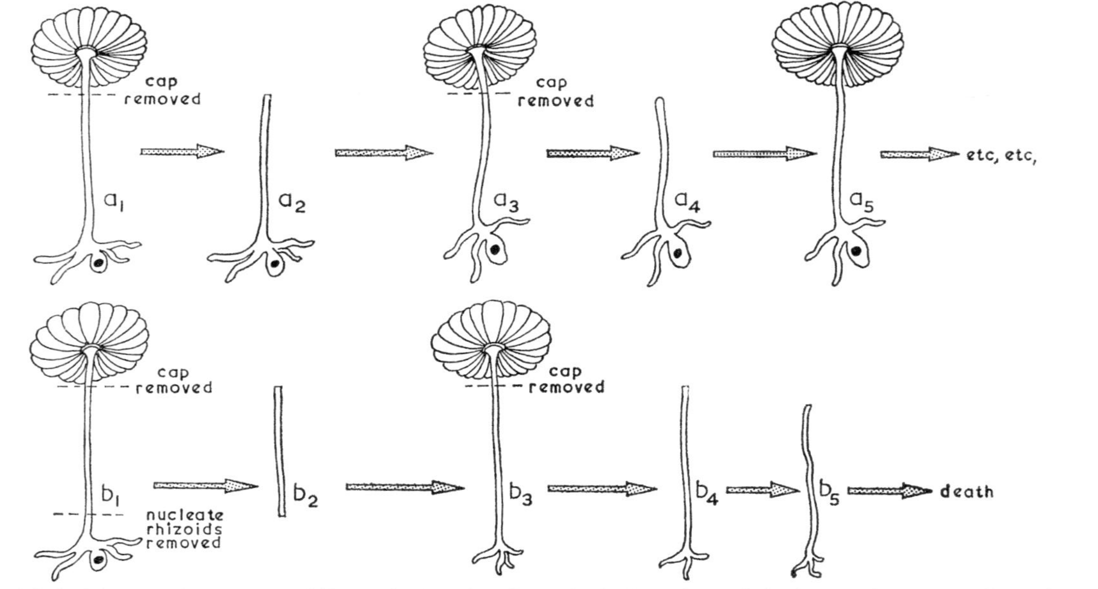

Fig. 10.4. The nucleus and differentiation in *Acetabularia*. The adult (a_1 and a_2 may be 50 mm long and consists of a stalk bearing rhizoids (with the nucleus) at the base and surmounted by a cap. This cap can be cut off but the nucleate stalk will form another; this can be done several times ($a_2 - a_5$). If the nucleus is removed, however, the capacity for cap formation is rapidly lost ($b_2 - b_5$). Note that weak, anucleate rhizoids may also be formed

because it is continuously formed by the new 'donor' nucleus (figure 10.5).

Experimental work on lower organisms (fungi and algae) demonstrates that the cytoplasm also can exercise a strong influence in differentiation. The activities of mitochondria and chloroplasts (which both contain DNA and RNA) play an important part, as do the 'plasmagenes' proposed by the geneticists.

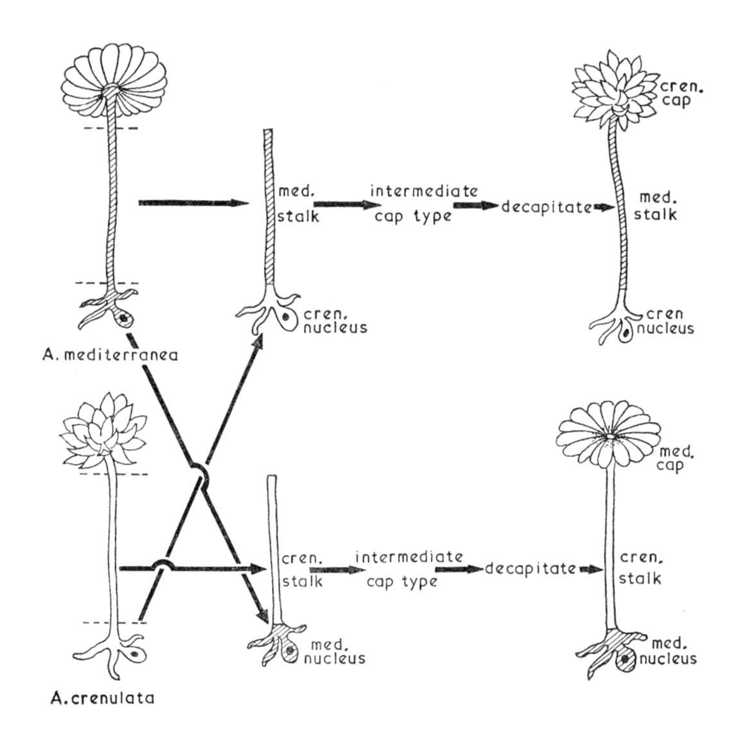

Fig. 10.5. The control of cap formation in *Acetabularia* by the nucleus. This illustrates how grafting experiments have been used to investigate the control exerted by the nucleus over differentiation. The characteristic of the cap is determined ultimately by the specific nucleus

Control at the molecular level

Genes

We know then that differentiation is under the influence of the nucleus. So instructions to the cytoplasm come from the nucleus, and we have seen in this and an earlier chapter that this is thought to occur in the guise of messenger RNA molecules. We have also seen that all nuclei probably have the full complement of genes necessary to make the organism, but obviously the set of genes which makes, say, a parenchyma cell must differ from that which makes a tracheid. We can infer, then, that some genes are 'switched off', i.e. they are prevented from making their specific mRNA and are therefore unable to pass instructions to the cytoplasm.

How is this done? A very ingenious idea was put forward a few years ago by the French scientists Jacob and Monod, on the basis of research done on bacteria. Certain substances fed to the cells can lead to the formation (*induction*) of particular enzyme proteins not produced in their absence. On the basis that the *potential* to form the enzyme must be present in the untreated cells, Jacob and Monod suggested that in the latter case the cells produce an internal *repressor* which prevents the gene in question from making the enzyme. The *inducer* acts, then, by combining with the endogenous repressor and rendering it inactive in its inhibitory action (*derepression* = induction).

Repressors themselves are formed by the action of special genes (regulatory genes) and in this way one gene can influence the activity of another.

The French research group found, furthermore, that some substances induce the formation of more than one enzyme, showing that several genes may be sensitive to a single endogenous repressor, and this has led to the hypothesis of the *operon* – a series of genes, often with related activities, which are all under the control of a single on–off switch on the chromosome called the *operator* (figure 10.6(*a*)). It is further possible that several operators may be sensitive to a single repressor substance, and thus a single compound supplied externally reacting with a single repressor

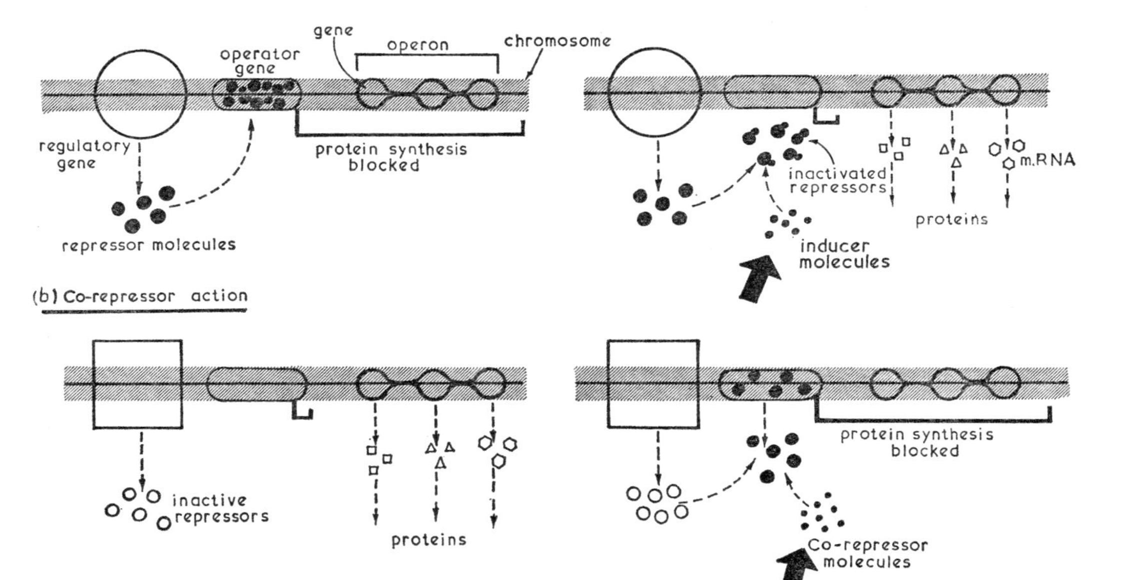

Fig. 10.6. Diagram of a current theory to explain the 'switching on' or 'switching off' of protein (enzyme) synthesis by particular genes. This theory is well established for bacteria, but has not been proved for higher plants

may result in the activity of many genes, leading to profound changes in the cell.

An alternative related situation also exists; it was found that some repressors are *inactive* unless specific compounds (*co-repressors*) are supplied. The inactive repressor combines with the co-repressor to give an active repressor (figure 10.6(*b*)). In such cases addition of the co-repressor *inhibits* the formation of the protein: an example of this type of action can be found in the effect of some amino acids to suppress the formation of certain enzymes in some bacteria.

The possible mechanism of the action of hormones is immediately obvious. They may be inducer substances which, entering the cell from other parts of the plant combine with repressors, thus activating operators; alternatively, in some cases their effect may be to act as co-repressors, switching off genes. The newly differentiating tissue so formed may now form hormones which act on other parts of the plant. Such controls could explain correlative growth.

Despite this attractive hypothesis, plant hormones have not yet been demonstrated to have this function, although it does seem (Chapter 8) that they do in some cases at least influence RNA synthesis.

But a further problem exists in complex organisms, which have a variety of cell types, which is not found in micro-organisms where every cell is like every other cell. Even assuming that hormones may be inducers or co-repressors, why do not, say, root apices form flowers when flowering hormone is induced in leaves treated with the right photo-period, whereas certain cells of stem apices do? The American plant physiologist Bonner, suggests that this is because parts of chromosomes are covered with certain proteins called *histones*. These proteins are known to be characteristic components of the chromosomes in nuclei. One kind of histone may 'cover' an operon maintaining it inactive even in the presence of the inducer, whereas another kind of histone covering the same operon in another nucleus allows the operon to be de-repressible.

Bonner based his idea on the results of experiments with pea chromosomes. He investigated the formation of a particular reserve protein which is formed in pea cotyledons during the development of the pea seed; it is not found elsewhere in the plant. His research team isolated chromosomes from tissue which had been ground up in such a way as to break open the cellular and nuclear membranes. The liberated chromosomes were then centrifuged away from the rest of the cellular components, and collected. These chromosomes are composed of *chromatin* which is a mixture of nucleic acid and protein. They are able to synthesise RNA (assumed to be messenger RNA) from a mixture of the four nucleotides which compose it (see Chapter 3). This RNA was able to promote protein synthesis in the test-tube when it was provided with ribosomes, amino acids, ATP and the necessary enzymes (Chapter 3). Now, chromatin made from parts of the plant other than the cotyledons did not lead to synthesis of the reserve-type protein, i.e. the corresponding gene was repressed, but chromatin made from cotyledons did, i.e. its gene was de-repressed. So Bonner had mirrored the living situation in this particular instance.

But why was histone thought to be involved in the repressed chromatin? By homogenising chromatin with dilute salt solutions it is possible to separate those parts of the DNA which are complexed with histone (nucleohistone). It was found that such material was hardly able to make RNA at all; on the other hand, DNA *separated* from histone by strong salt treatment *was* able to make RNA. So histone appears able to block gene expression by preventing mRNA formation.

The last experiment of this series showed that the gene for synthesis of the reserve-type protein is, in fact, present in apical bud chromatin (which, of course, is unable to make the protein), i.e. *histone-free* DNA from this chromatin was able to make the RNA necessary to make the reserve protein (Table 3).

Bonner's results shown diagrammatically in figure 10.7 were not as clearcut as is required to prove a hypothesis, and there is some doubt, therefore, about the argument given here. They do show, nevertheless, how research workers are tackling the important problem of gene expression.

Enzymes

A completely different control, occurring in the cytoplasm, is *end product inhibition* which can affect some biosynthetic reactions. The research which elucidated this process was also done using bacteria.

When certain compounds (usually amino acids) which are synthesised by the cells were added externally to excess, the biosynthetic pathway for the particular compound ceased acting. When the amino acid provided had been used up during growth, synthesis started, showing the control to be reversible.

It was found that the amino acid specifically inhibits the action of one of the enzymes involved in the biosynthetic sequence, by changing its shape and thereby preventing it from combining with its substrate – the so-called *allosteric* effect. When the amino

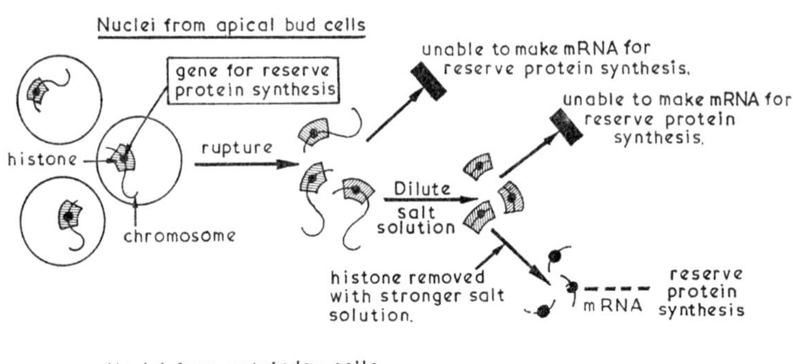

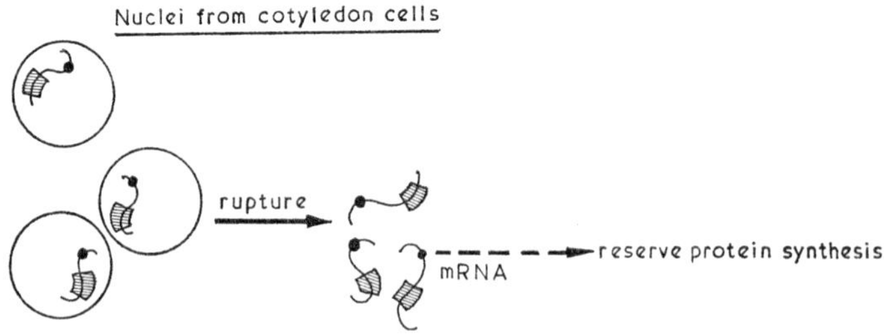

Fig. 10.7. Diagram of Bonner's experiment to demonstrate the action of the protein histone in repressing gene action. It is assumed that all nuclei in the plant contain chromosomes with the gene for reserve protein synthesis, but only those from cotyledon cells do, in fact, make reserve protein. The experiment was designed to demonstrate that histone 'covers' (represses) the gene in other cells

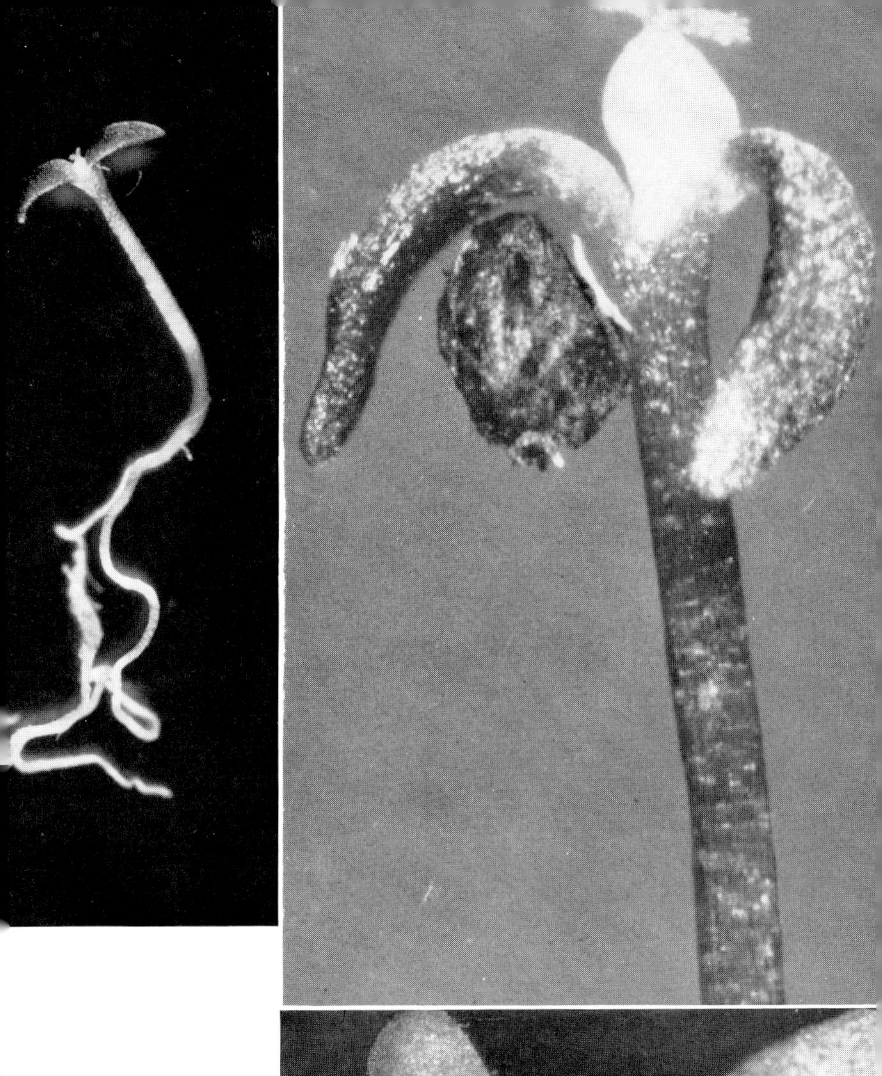

Plate 15 *Flower formation in seedlings of* Chenopodium. This species is extremely sensitive to photoperiod and forms rudimentary flowers even in the very young seedling stage, before mature leaves form.

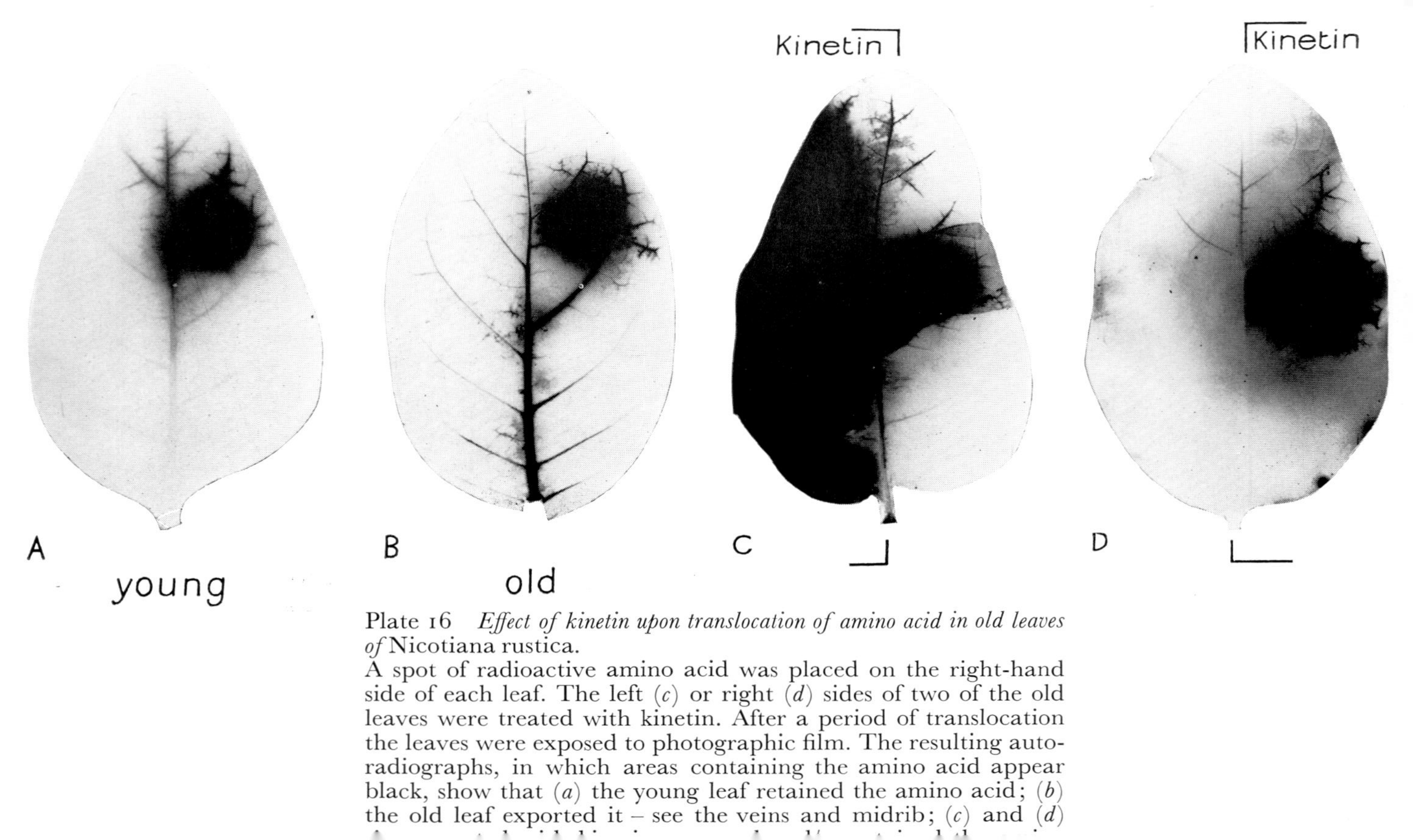

Plate 16 *Effect of kinetin upon translocation of amino acid in old leaves of* Nicotiana rustica.
A spot of radioactive amino acid was placed on the right-hand side of each leaf. The left (*c*) or right (*d*) sides of two of the old leaves were treated with kinetin. After a period of translocation the leaves were exposed to photographic film. The resulting autoradiographs, in which areas containing the amino acid appear black, show that (*a*) the young leaf retained the amino acid; (*b*) the old leaf exported it – see the veins and midrib; (*c*) and (*d*)

acid is removed, the enzyme reverts to its original active shape. In this way the synthesis of just the right amount of compounds required for growth may be controlled, the synthetic sequence being switched off by sufficient end product.

Table 3

Synthesis of pea-seed reserve protein

	Reserve protein synthesis (arbitrary units)
DNA from chromatin of cotyledon cells (control)	33
Whole chromatin from apical bud cells (i.e. histone present)	0
DNA from apical bud chromatin (i.e. histone removed)	32

(recalculated from Bonner, *The Molecular Biology of Development* Clarendon Press, 1965 *p.* 23)

But once again the significance of this sort of control in plant cells is not yet proved.

We can be sure that there are many other types of control of cell development of which we are still unaware. Discovering them is one of the most challenging and rewarding research topics in biology at the present time.

Summary

In this chapter we have explored the internal factors which may be involved in controlling the development of different types of cells, tissues and organs. Certainly the nucleus exerts an overall influence, but the information which it transmits to the rest of the cell is affected markedly by hormones, which come into the cell from other parts of the plant.

Although we know what these hormones are in some cases, e.g. auxins, gibberellins and cytokinins, and can demonstrate some of the effects they have on development, we can only speculate on their mechanism and action. We naturally look to theories of gene action as a source of ideas, and we find that hormones may be inducers or de-repressors or co-repressors in switching certain genes on or off via operator genes.

The additional problem of why a hormone might affect the nucleus of one type of cell, e.g. from a shoot apex, while not affecting another, e.g. from a root apex, may be explained by the protection, or covering over, of some genes by special proteins, histones, which are found in the nucleus associated with chromosomes.

We must not lose sight, also, of the controlling factors which operate in the cytoplasm, such as concentrations of end products causing end product inhibition of enzyme activity: there are many more of these, most of them as yet unrecognised.

11

Flowering

With very few exceptions almost all angiosperms flower sometime during their life. In the so-called monocarpic plants this occurs only once, either during their first or second year (annuals and biennials) or after many years of purely vegetative growth. The bamboos and the century plant are examples of the latter and, like the other monocarpic plants, their life is terminated by flowering. The polycarpic plants (perennials) flower often and survive for many years as woody plants or as perennating organs (bulbs, rhizomes, etc.).

The normal vegetative apical meristem of a shoot is indeterminate, i.e. it continues to produce primordia for an indefinite period. Theoretically, this need never stop, and in trees we find apical meristems which have been active for many years although, of course, they undergo intermittent (usually annual) periods of dormancy. When an apex develops into a flower, it is given over entirely to the formation of flower primordia and is then incapable of further morphogenetic activity, i.e. it becomes a determinate apex. How one kind of apex is converted to the other is a fascinating morphogenetic problem, of the kind which lies at the core of biology. We do not understand the physiology and biochemistry of this transformation but we will see later that it is induced by signals coming from the leaves.

Since the flowers of most plants are bisexual, primordia of stamens and pistils are formed during the conversion of an apex. However, some species have unisexual flowers, bearing either stamens or pistils. A further control, one which decides sex, therefore operates at the apex. In some cases, daylength and temperature are involved in this control. For example, long days and high temperature favour the production of staminate flowers in one species of *Cucurbita*. Internal controls, the hormones, can also determine sex expression. High auxin levels generally promote femaleness whereas gibberellins induce the formation of male flowers. In fact, the future sex of some plants can be controlled by the application of the appropriate hormone.

Even to a casual observer it is noticeable that flowering in different plants of the same species occurs at the same time. Biologically, this is extremely important as many species must be cross-fertilised; even when this is not a necessity it is nevertheless of considerable genetic advantage.

The synchronisation of flowering suggests the existence of some precise control factors. The temperature and the daylength are the important environmental ones. However, in many plants flowering is not under environmental control and here endogenous processes are at work. Factors such as actual age and physiological age (e.g. number of nodes produced) are especially important. Most scientific work on flowering has been concerned with control by daylength and temperature which we will now consider in further detail.

The photoperiodic control of flowering

Floral induction

The control of flowering by the daylength was called photoperiodism by Garner and Allard who performed the first systematic experiments. We now appreciate that many developmental phases in plants are regulated by the photoperiod (i.e. the daylength) as so the term is no longer restricted to flowering.

Garner and Allard worked with the varieties of tobacco and soya-bean, Maryland Mammoth and Biloxi respectively. They found that Maryland Mammoth grown in the fields in summer reached over 10 feet without flowering but in the greenhouse in winter formed flowers when only 5 feet high. Biloxi soya-bean always flowered in September whether the plants were 60 or 120 days old. It was soon recognised that daylength was the important controlling factor and that both these plants required short days for flower induction.

The terms short-day plants (SDP), long-day plants (LDP), and day-neutral plants were coined to describe the response types first discovered. Now, however, we know of more than three photoperiodic classes. Some plants, for example, require short days and then long days; these are the short–long-day plants. A recent monograph on flowering lists forty-six categories of photoperiodic plants, many of which exhibit complex interactions between temperature and daylength.

Short-day plants and LDP are those whose flowering is promoted by daily photoperiods respectively shorter or longer than a certain value, called the critical daylength. Some species of *Chrysanthemum*, *Xanthium pennsylvanicum* (cocklebur) and *Chenopodium* (goosefoot) are examples of short-day plants. Spinach, radish, *Hyoscyamus niger* (henbane) and oats are all LDP. Day-neutral plants include tomato, garden peas, broad-beans and pansy (Plate 14).

Many species, such as *Xanthium* and *Hyoscyamus* have a strict requirement for the correct daylength and do not flower if they are exposed to days which are too long or too short by only 30 minutes (figure 11.1)! This qualitative response indicates that some plants can measure time amazingly accurately. Even some tropical plants, which live at latitudes where over the whole year variation in daylength is 50 minutes, flower only in September when the daylength just falls below the critical 12¼ hours. Figure 11.1 also gives a curve for flowering response which does not show such a sharp critical daylength. This is a quantitative response to daylength.

We should be clear that in our definitions of SDP and LDP no absolute daylength values are suggested since there is so much variation among different species. Some SDP will flower under exactly the same daylength as some LDP; however, decreasing daylength continues to promote the former but inhibits the latter (figure 11.1).

Although plants may have fairly rigorous photoperiodic requirements for flower formation they do not need to be maintained continuously in the inductive daylength. *Xanthium*, for example, is extremely sensitive and only one short day, of less than 15½ hours photoperiod, is sufficient to induce flower formation, which is visible two or three weeks later. Most other species, however, need to be exposed to several inductive day-lengths before they will later form flowers.

Having seen that plants respond to the daylength, we may now pose two questions. How do plants 'know' what the daylength is, and, when they do know, how do they use the information? The information is almost certainly used to make a flowering hormone. We will now see what evidence there is for this hormone and where it is made.

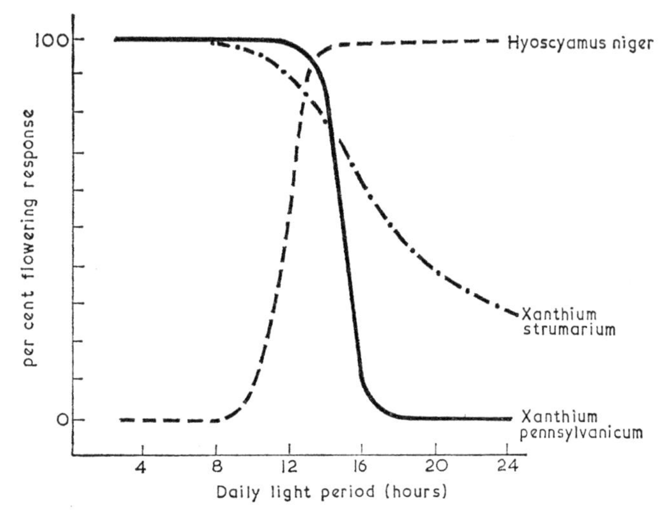

Fig. 11.1. The effect of daylength on the flowering response of 3 different photoperiodic categories—*Hyoscyamus niger*, a long-day plant, *Xanthium pennsylvanicum*, a short-day plant, and *Xanthium strumarium*, a short-day plant with a quantitative response. Note the sharp critical daylength for *Hyoscyamus* and *X. pennsylvanicum*. Note also that at about 13 hours light per day both of these will flower

Flowering hormones

Shoot apices respond to the daylength by forming flowers; but do they actually perceive the stimulus? Experiments of the kind illustrated in figure 11.2 show that perception of the photoperiod occurs in the leaves, and the youngest mature leaves are generally

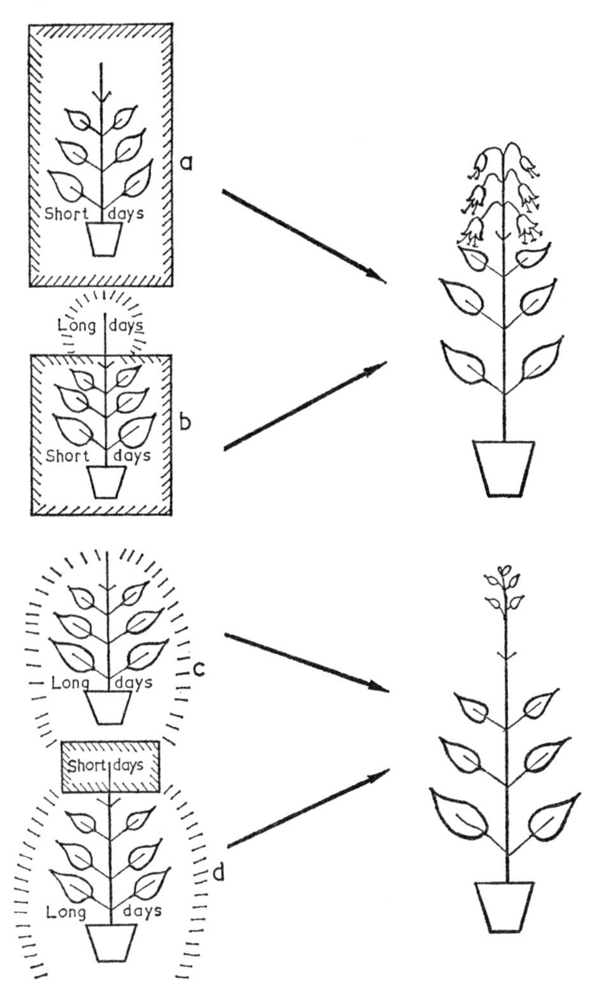

Fig. 11.2. Perception of the correct day length is by the young, mature leaves. This is a short-day plant. The very young leaves are removed so that the apex (which forms the flowers) or the maturer leaves can be exposed separately to the inductive day length (i.e. short days). Note that only when the leaves are given short days does flowering subsequently occur (*a* and *b*).

most sensitive. *Chenopodium* and *Pharbitis* though, are sensitive in the cotyledon stage, and in the former, rudimentary flowers can even be induced in a seedling bearing no mature leaves (Plate 15).

In the light of these findings we begin to think in terms of a transmissible influence from the leaves; in short, a flowering hormone. Grafting experiments provide very good evidence for such a hormone, sometimes called florigen. A photoperiodically-induced *Xanthium* will promote flowering in a non-induced plant when the two are grafted together. Florigen, it seems, can travel from stock to scion provided there is good phloem connection. Even isolated leaves can be photoperiodically induced and act as a 'donors' after grafting. Perhaps the most fascinating experiments are those where a LDP is grafted on to a closely-related SDP. For example, when the LDP *Hyoscyamus* is grafted on to the SDP Maryland Mammoth tobacco (both in the *Solanaceae*), the latter induces the former to flower when the union is kept in short days! The converse induction is also possible, and even day-neutral plants can promote flowering in photoperiodic species. The flowering hormone seems, therefore, to be physiologically identical in all categories of plant. But this conclusion raises the intriguing problem of why some plants can synthesise the hormone only under short days, some others under long days and yet others in any daylength.

So far, there has been only marginal success in isolating the hormones and since we do not know what they are chemically we cannot say how far their equivalence really extends. We might ask, though, how they are related to the known plant hormones, described in Chapter 4. The flowering hormone is not an auxin, for the following reasons. Consistent changes in auxin levels in response to inductive daylength do not occur. Secondly, except in one 'classical' exception – the pineapple – flowering cannot be induced by applying auxin. In fact, auxin tends to inhibit flowering in SDP. Gibberellins, though, cause flowering when applied to many LDP – those which form rosettes in short days. Moreover, the gibberellin level in some LDP (e.g. spinach) rises in long days. Nevertheless, gibberellin cannot be the flowering hormone, particularly because it does not induce flowering in SDP.

Whatever its nature, the flowering stimulus causes profound biochemical changes at the apex. We know little about these but, as we might expect, the synthesis of protein and nucleic acids are involved.

We have so far considered promotive hormonal actions in flowering. However, there is evidence that 'incorrect' daylengths might actively inhibit flowering. This seems to be the case in some SDP (e.g. strawberry) where an inhibitory effect of long days can be transmitted along the runner from one plant to another. This is difficult to reconcile with the florigen concept and we do not yet know if it is of general significance.

The synthesis of flowering hormone results, then, when leaves have measured the appropriate daylength. Now we have to see how this perception occurs.

Measurement of daylength

The daylength is composed of two factors – the light and dark periods. In photoperiodic induction both or only one may be important. In a cycle of fixed length (e.g. 24 hours) variation in the length of the photoperiod is inevitably accompanied by variation in the dark period. But with suitable experimental techniques light and dark regimes having only one variable can be obtained; thus the relative importance of light and dark can be assessed. It is then found that the *absolute* length of the dark period is the basis of the photoperiodic stimulus, not the light period or the ratio of light to dark. For example, the SDP *Xanthium* is induced by 16-hour dark periods even when the accompanying photoperiods are 20 hours long. In qualitative SDP and LDP the dark period must therefore be respectively greater or less than a certain critical minimum. The dark period is clearly favourable to flowering in SDP but not in LDP (in fact, the latter do not require darkness). Perhaps we should properly refer, therefore, to long- and short-night plants. Provided the dark period is of the right length, then, flowering is induced. So now the problem is, how does the plant measure darkness? A long dark period is completely nullified when it is interrupted in the middle by a short (say 30 min.) 'light-break'. Flowering in LDP is then promoted but it is inhibited in SDP (figure 11.3).

The light energies involved in the effect are very low and clearly the 'clock' for measuring the length of the dark period is a light-sensitive one. The action spectrum for the 'light-break' shows that red light is responsible and that it is reversible by far-red. This is characteristic of the phytochrome system (Chapter 9), so this pigment is involved in the timing mechanism.

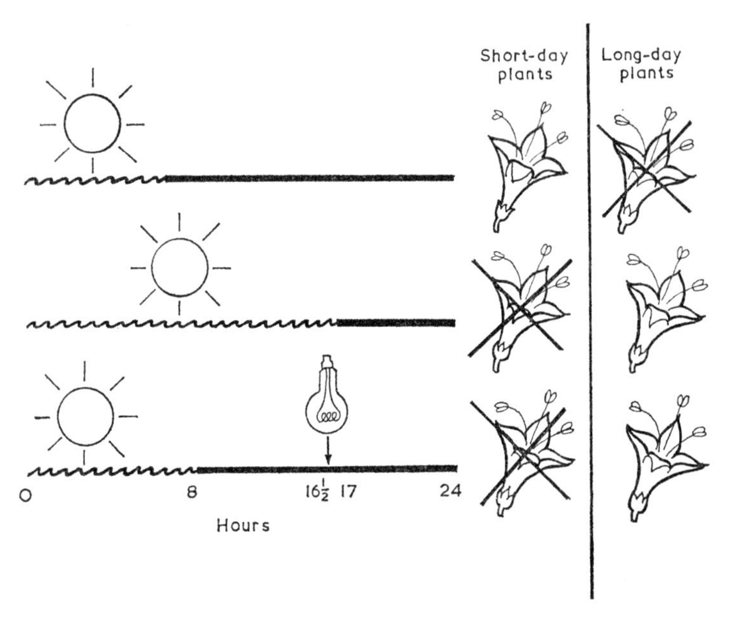

Fig. 11.3. The importance of the absolute length of the dark period as illustrated by the effects of a 'light-break'. Short-day plants flower when the dark period (straight line) is sufficiently long. Long-day plants flower when it is sufficiently short. A long dark period is nullified by a 30 minute light break; flowering is thus prevented in SDP but promoted in LDP

Phytochrome may function in timing in the following way. At the end of the main photoperiod phytochrome exists as P_{730}, which, we can postulate, inhibits 'florigen' synthesis in SDP but promotes it in LDP. Now in the subsequent dark period slow reversion of P_{730} to P_{660} occurs (page 120). In SDP, the dark period must be long enough to allow this reversion to go to completion (so that florigen synthesis can then take place) but in LDP it must be prevented. This is an attractive hypothesis which would explain the effect of 'light-breaks' but unfortunately it does not seem quite valid. Far-red given at the end of the main photoperiod should greatly shorten the critical dark period in SDP by causing a rapid conversion of P_{730} to P_{660} but in fact it only shortens it by an hour or so. Similarly the same far-red should inhibit induction in LDP but it generally does not.

Another approach to the timing mechanism is based on endogenous rhythms. Rhythmic behaviour in plants is manifest, for example, in daily leaf movements. Similar metabolic rhythms, of unknown biochemistry, may be important in photoperiodism. There are thought to be two phases in the 24-hour oscillation, one in which light is 'favourable' to flower induction (photophile) and one where it is not (scotophile). SDP and LDP both have similar rhythms but in SDP it begins immediately the plant is illuminated whereas in LDP there is a delay of 8–12 hours. As shown in figure 11.4 this means that the phase requirements (i.e. light in the photophile phase) in SDP are satisfied only by short days and conversely for LDP. More evidence is now accumulating for this hypothesis but the nature of the rhythm and how phytochrome interacts with it are both unknown.

We have emphasised the roles of the dark period and the phytochrome system. However, it seems that in some SDP the long dark period must also be preceded and followed by high-intensity light. This seems unconnected with phytochrome; it may partially fulfil a photosynthetic requirement but this is not certain.

In summary we may say that flowering is induced in SDP and LDP by dark periods greater or less than a certain length. The duration of the dark period is measured by a mechanism in which phytochrome is involved. Once the dark period has been satisfactorily timed, synthesis of the floral stimulus (florigen) begins in the leaves, and it is then translocated to the apex in which morphogenetic changes are induced.

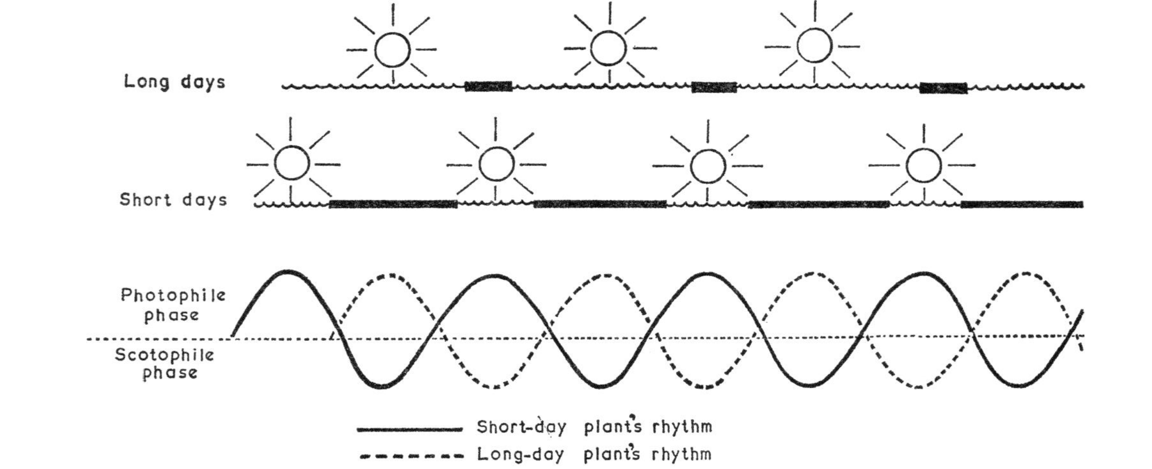

Fig. 11.4. A possible operation of endogenous rhythms in the control of flowering. It can be seen that in SDP, where the rhythm begins immediately, the phase requirements for light and darkness (i.e. light in the photophile phase, dark in the scotophile) are satisfied only by short days. In LDP the rhythm (although the same) starts later, and is thus satisfied only by long days

Vernalisation

Low temperature treatments also promote subsequent flowering in many species. This is called vernalisation. Many biennials and perennials must experience some time at low temperatures for flowering subsequently to occur at normal temperatures (e.g. cabbage, celery). Hence, some plants are restricted to climates in which the winters are fairly cold.

In many plants chilling is effective in the seed stage. Winter wheat and rye, for example, are sown in the autumn and become vernalised during the winter. Temperatures from 0–5°C are most effective but their action is erased when warm temperatures follow immediately (devernalisation).

The spring strain of rye is a quantitative LDP and in long days it forms flowers after 7 leaves have been produced; in short days, however, 22 leaves are first formed. If the winter strain is unvernalised it behaves as day-neutral and flowers after 22 leaves have formed, no matter what the daylength. After vernalisation, though, it is a quantitative LDP and is then like the spring strain.

The cold treatment is effective on the shoot apex, in both the embryo (e.g. cereals) and in adult plants (e.g. cabbage). There is no consistent evidence for the formation of a mobile 'substance' during vernalisation. All growth arising from a previously chilled apex behaves as if vernalised, while non-chilled apices on the same plant remain unstimulated. Essentially what vernalisation does, at least in rye, is to convert the apex so that it becomes sensitive to the long-day 'signals' from the leaves.

The mechanism of vernalisation is not understood but it seems possible that gibberellins are involved. These hormones replace the cold treatment in many biennials (Plate 8), and there is some evidence to show that chilling may lead to increased endogenous gibberellins.

Thus, the environment regulates reproduction. Profound changes in the plant are brought about by daylength and temperature and through these changes reproductive behaviour is synchronised.

Summary

Flowering in some plants is determined apparently entirely by internal factors, but in very many species the daylength (photoperiod) or temperature plays a profound role.

In plants of the various photoperiodic categories it seems that flowering hormones ('florigen') are synthesised in the leaves which are translocated to the apical meristem; this is then converted into a flower-forming meristem. Hormone synthesis can begin only when the 'correct daylength' (or, more accurately, dark period) has been measured. This measurement involves a 'clock' which utilises phytochrome possibly together with endogenous rhythms whose metabolic features are unknown.

Low temperature treatment (vernalisation) leads to subsequent flower formation in many plants, some of which are sensitive in the seed stage (e.g. cereals). Here, vernalisation renders the plant sensitive to photoperiod.

Plant hormones, especially gibberellins, are implicated in flowering in long-day plants and in vernalisation, but their general role in flowering is uncertain.

12

Dormancy

In all plants (at least in temperate climates) there is a period of dormancy when growth is suspended and metabolism is at an extremely low ebb. One form in which plants pass through this phase is as seeds. Many species do not have any other dormant organs and the whole plant dies after reproduction and seed formation is over. Other species, though, survive as dormant buds on woody shoots, tubers, rhizomes and similar organs, or as bulbs and corms.

This phenomenon points to the intriguing question of how life can be maintained in cells which are relatively very dry and in which metabolic activity is barely measurable. The problem is posed very forcefully by those seeds which have remained alive (viable) for many hundreds of years. Some of the oldest surviving seeds are those of *Nelumbo nucifera* (Indian Lotus) which are almost certainly 1200 years old. The record, though, is claimed for a species of *Lupinus* whose seeds, found in the Yukon, are believed to be 10,000 years old!

Often, growth of 'dormant' organs (e.g. pea or bean seeds) is resumed upon reabsorption of water at a suitable physiological temperature and, in most cases, an adequate air supply. On the other hand, many seeds and buds fail to grow even then. In fact, we shall define dormancy as that state of a tissue or organ such that it will not resume growth even when all normal conditions (e.g. temperature, air, water) are satisfied. Some special internal and/or environmental control, must, therefore, exist. This invariably means that a 'trigger' is required to release the organ from its dormancy. We will soon consider what these triggers are.

Dormancy is extremely important to the plant as it provides a means of survival, in spite of inclement external conditions.

Dry dormant seeds can withstand freezing, fairly high temperatures, and require neither light nor air. Buds on woody twigs survive through periods of extreme cold, of drought, and the short dark days of winter. Moreover, dormancy sometimes confers upon the plant a dispersion in time comparable in biological importance with the spatial distribution of seeds and fruits effected by animals, wind and other means. An example is the unsynchronised breaking of dormancy in a seed population, which then produces seedlings intermittently over a period of time. Certain weed species demonstrate this, and the chances of survival of the species are thereby strengthened in spite of annual attacks on adult plants by the farmer or gardener! We will also see that devices are built into the dormancy mechanism by which plants can 'sense' their environment. For example, seeds of some desert plants can 'measure' the amount of rainfall; and buds can detect the difference between the warmth of autumn when they do not grow, and that of spring, when they do.

The onset of dormancy

What causes a shoot apex to stop growing and form a dormant bud, or a seed to develop such dormancy that some special condition is later necessary to induce germination?

In many woody and herbaceous species endogenous, possibly rhythmic processes, are involved, which are, as yet, but little understood. But in very many other cases environmental factors, particularly temperature and daylength, are important. For example, relatively high temperatures, combined with a lowered oxygen supply (caused by the tightly-packed leaves around the apex), might be the cause of bud dormancy in some plants.

The role of the daylength has been studied extensively and we know of many species (e.g. birch, beech and sycamore) in which short days cause apical growth to stop and dormant buds to form. This can be prevented by long days, in some species for quite a long time. *Robinia pseudacacia*, for example, grows continuously for at least 18 months in artificially arranged long days. Short days also affect herbaceous species, and induce tuber formation in *Begonia* and potato. On the other hand, the formation of bulbs in onion is a long-day response. Less is known about the effect of the environment on the induction of seed dormancy, though it has been claimed that daylength conditions during seed maturation can have an effect. Mature, but non-dormant seeds of some species (e.g. *Nemophila*) can be made very dormant by long-day treatments. Since this is an effect on an already existing seed it is called secondary dormancy.

Just as in the control of flowering the perception of the photoperiod is by the leaves, and these organs, not the apices, must be exposed to short days before apical growth stops. An apparent exception is birch, but here the young unfolded leaves in the apical bud are probably more sensitive than the mature ones. Presumably, some 'influence' must be transmitted from the leaves to an apical meristem to make it dormant. There is good evidence that in certain woody species this 'influence' is an inhibitor. For example, there is a great increase in the inhibitor content of sycamore leaves, kept in short days, followed later by a similar increase in the growing point. Very good experimental confirmation of the idea that inhibitors transform a growing apex into a dormant bud has been obtained in the case of birch. The leaves were exposed to short days and the inhibitor extracted. It was then fed through the leaves into another shoot kept in long days. The apex of this shoot then stopped growing and formed a dormant bud. Now, the inhibitor in birch seems to be very similar to the one from sycamore. The latter has been purified and was first called dormin, but now is known as abscisic acid (page 63).

It appears, therefore, that photoperiodically-induced bud dormancy, at least in some woody species, is caused by an inhibitor. Dormancy, once induced, must be maintained and presumably the inhibitor also does this. We will now consider how inhibitors and other factors are involved in the general maintenance of dormancy of seeds and buds and how these factors are overcome to allow of renewed growth.

The maintenance and breaking of dormancy

The problems we now have to consider are these: why do plant organs remain dormant even under apparently favourable growing conditions, and what eventually enables them to resume growth? As a general guide, we may say that dormancy is maintained by some internal control mechanism. The organ alone sometimes gradually escapes from this control (a process called after-ripening) but more often changes induced by environmental factors such as light or low temperature are necessary. This implies that the maintenance and breaking of dormancy depend on an interaction between the environment and the plant. Before discussing how this works we will deal with the internal control systems.

Internal controls

The best known internal mechanisms for maintaining dormancy are inhibitors and the structures enclosing the embryo or apex. Occasionally, other important features are encountered, such as immaturity of the embryo, but these details do not concern us now.

Inhibitors

These have been found in very many dormant tissues. They are present in potato buds (and in the peel) in ash buds and, as we saw above, in birch and sycamore buds, to give but a few examples. Ammonia, cyanide, coumarin, abscisic acid and a host of other inhibitory substances, many still unidentified chemically, occur in a variety of seeds. In *Iris* seeds, a powerful inhibitor diffuses from the endosperm on to moist filter-paper and prevents any other seeds there from germinating. Seeds of many desert plants are so rich in water-soluble inhibitors that germination proceeds only when these are washed out. In nature, this occurs in really heavy but not light rainfalls. Thus, the seeds 'measure' the rainfall and a seedling is produced only when the soil has sufficient water to support growth for some time. This, incidentally, is a simple example of how environment interacts with an internal factor.

Only in a few cases (e.g. ash and potato buds) do the inhibitors in dormant tissues decline prior to the resumption of growth, and so the natural breakdown or inactivation of inhibitor is generally not involved in the loss of dormancy. Clearly some other change must occur which enables the seed or bud to overcome the effects of these substances.

The enclosing structures

The 'seed' coat (testa or pericarp) is often involved in maintaining dormancy as we can see from the fact that embryos of many dormant seeds germinate when isolated (e.g. lettuce, birch). In some cases, this treatment allows inhibitors to diffuse out or it permits more oxygen to reach the embryo, or carbon dioxide (a narcotic) to escape. The extra oxygen is sometimes used by the tissues of the embryo to achieve the enzymic oxidation of inhibitors (e.g. in *Xanthium*). Under natural conditions, though, seed coats rarely get removed, and we find that their effects are overcome only by increasing the growth potential of the embryo. This itself is often brought about by some environmental 'trigger'. Let us now see what these environmental controls are.

Environmental controls

Temperature. Buds on woody twigs of many species, when kept in the greenhouse over the winter, are still dormant in springtime, even when buds left outside have resumed growth. This is because a period at temperatures just above freezing ('chilling') is required for the subsequent breaking of dormancy. This is found in poplar, ash and sycamore. Other dormant buds, such as those on potato tubers, respond similarly to low temperatures. Chilling is also effective in very many seeds, for example, peach, apple, ash and hazel. The need for chilling is of undoubted ecological importance since it ensures that bud growth or seed germination is prevented until after the winter has passed. Further, it is partly responsible for limiting certain plants to climatic regions which have a relatively cold winter.

Light. The importance of light in the breaking of dormancy is dramatically evident in those seeds which are stimulated to germinate by illumination, perhaps for only a few seconds or minutes (e.g. tobacco, some varieties of lettuce). Red light is most effective and, in fact, the phytochrome system is involved. Many seeds have this natural sensitivity to light but other, light-insensitive seeds can be made light-requiring by treatment with the inhibitors coumarin and naringenin. This suggests that the normal possession of a light-requirement might be induced by endogenous inhibitors, and indeed these do occur in some such seeds.

Some seeds require repeated daily exposure to light. Birch seeds, for example, are photoperiodic and require long days before they will germinate. In striking similarity unchilled birch buds must also be exposed to long days before they commence growth. Long days also break bud dormancy in beech and accelerate bud break in poplar and sycamore. We can see, therefore, that photoperiodic conditions break dormancy as well as induce it (page 166).

Interactions between factors

In the examples just mentioned – poplar and sycamore – long days only accelerate bud break but they do not initiate it; this is achieved by chilling. We see in these two cases examples of how environmental factors can interact. In fact, such interactions are found in very many seeds and buds, often in a more complicated form. Often, no single environmental factor is required to break dormancy but any one of a number will do. Two examples from the same species illustrate this point. Intact, unchilled birch seeds remain dormant at 15 °C unless exposed to about eight consecutive long days. At 20 °C, however, one single 5-minute exposure to light is sufficient to cause germination, which begins a few days after illumination. In contrast, embryos taken out of the seeds germinate in complete darkness. Washing whole seeds (which removes inhibitors) or treatment with high oxygen concentrations, both enhance germination even in short days. Chilling causes dramatic changes, for then whole seeds behave like isolated embryos – they germinate in darkness! Similarly, unchilled birch buds require long days to resume their growth, but chilled buds will do this even in complete darkness. These examples indicate the complexity of interactions between the various controls. The parallels between birch seeds and buds strongly suggest that the control mechanisms for dormancy of these different organs are basically similar.

How do environmental controls act?

We might speculate that chilling or light could bring about a change in the internal controls, for example inhibitors. As we have already mentioned, though, a drop in inhibitor level has only very rarely been recorded, and it does not seem, therefore, that environmental factors act in this way. But the growth potential of the embryo, or the bud, is evidently enhanced; how else might this be achieved?

In growth processes we expect phytohormones to be involved, and it is possible, then, that they are implicated in the breaking of dormancy. The auxins are apparently not concerned; they have very little effect on dormant seeds or buds, nor do they show any concentration changes associated with dormancy breaking. However, growth promoters are certainly involved in dormancy, especially the gibberellins. Chilling causes an appreciable rise in endogenous gibberellins in ash and hazel seeds (figure 12.1) and in sycamore buds, where build-up of these hormones precedes bud break. Thus, one general action of chilling may well be to increase internal gibberellin levels, which then overcome the inhibitors that are present. Indeed, this type of hormonal interaction has been achieved experimentally, for the dormant birch buds which were induced to form by inhibitor treatments (page 167) resumed growth when treated with gibberellic acid. Applied gibberellins, in fact, strongly affect dormant organs and many seeds and buds are stimulated to grow by these chemicals. Furthermore, the onset of dormancy in some instances (wild oat seeds and potato tuber buds) can be prevented by treating the parent plants with gibberellic acid during seed or tuber maturation. The slow loss of dormancy in wild oat seeds which occurs over a year or so is associated with an increased capacity of the seed to synthesise gibberellins.

The importance of the gibberellins in the breaking of dormancy is, then, well established, but it is too soon to say whether other treatments besides chilling (such as light) act through the intermediary of these hormones. Moreover, even the beneficial effect of chilling is often localised just to the branch which was exposed to low temperature – an unexpected phenomenon if a transmissible hormone is induced. So although the evidence implicating gibberellins in dormancy is very strong, these hormones do not provide us with the complete answer.

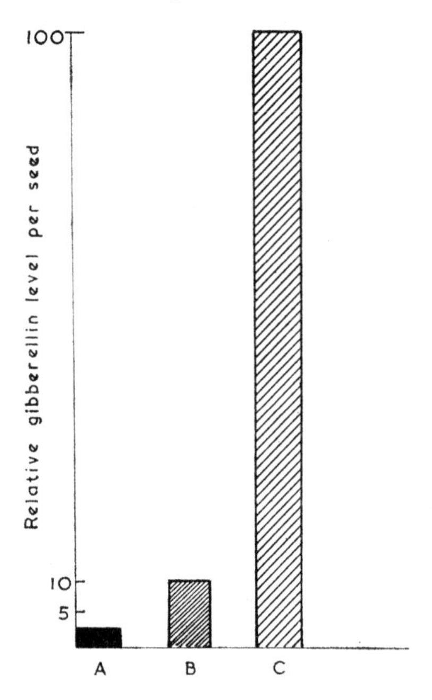

Fig. 12.1. The change in the level of extractable gibberellins of hazel seeds after chilling.
A = seeds extracted immediately after 28 days at 5 °C
B = seeds extracted after 28 days at 5 °C, followed by 2 days at 20 °C
C = seeds extracted after 28 days at 5 °C, followed by 7 days at 20 °C

We will conclude this discussion of the action of environmental controls by turning to a basic cellular process – RNA synthesis. When chromatin (i.e. DNA and associated protein) is isolated from potato buds which have been given a dormancy-breaking treatment, it will support RNA synthesis in a test-tube, provided certain enzymes and nucleotides (page 41) are present. However, chromatin taken from dormant buds cannot do this. Thus, DNA from the dormant tissue is apparently suppressed. This raises the fascinating possibility that the interactions between the environment and the internal control systems which we have outlined above may operate fundamentally at nuclear level, albeit through various hormones, and the success of an environmental factor in breaking dormancy is a reflection of its ability to act, indirectly, as a gene 'switch'.

Summary

Dormant buds on woody twigs and other organs, and seeds are the means by which plants can resist inclement environmental conditions and by which they can be distributed in time.

The term 'dormancy' is taken to mean that some special factor is required for the resumption of growth. Often, these are environmental; sometimes internal changes must occur (e.g. after-ripening). Light (especially daylength) is involved in the onset of dormancy and, together with low temperature, also in the breaking of dormancy, in both seeds and buds; this suggests that the mechanism of dormancy in these organs is similar.

Environmental factors are known in some cases to work through endogenous hormones, e.g. inhibitors and gibberellins. Details concerning the control mechanisms are known only in a few cases. Even less is known about the basic cellular control but the on–off operation of gene switches may be especially important in dormancy.

13

Senescence

As far as we know all organisms eventually die. If we exclude sudden death due to accidents, death is preceded by a period of gradual decline called senescence.

Higher plants differ from animals in their senescence. Often individual organs such as leaves, flowers or fruits may senesce and die while the rest of the plant is healthy and growing, and all stages from vigorous growth at the apex of a shoot to senescence

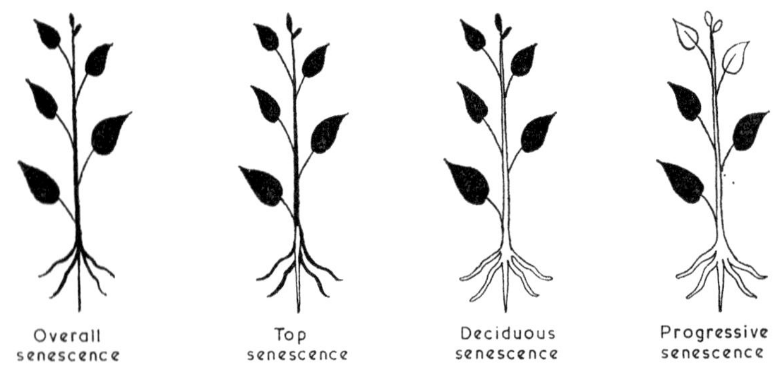

Fig. 13.1. Patterns of senescence shown by plants. (after Leopold)

and death at the base can commonly be seen on the same plant. Figure 13.1 shows the different patterns of senescence which occur in plants.

During the process, important nutrient elements such as nitrogen, in the form of amino acids derived from the break-down of proteins, may be withdrawn from senescing leaves and translocated to the growing regions; in some cases at least this directional flow of nutrients towards the growing points may itself be the cause of senescence of the exporting organs. Broad-bean cotyledons show this well: if the growing shoot is cut off above the cotyledons, they stay green and healthy when otherwise they would have yellowed and died.

Senescence is usually a functional part of the life-cycle. For plants with a specifically limited lifespan, i.e. annuals and biennials, it is often intimately associated with the ripening process of fruits and seeds; this is shown very strikingly by cereals in which the whole plant senesces as the ears ripen. Senescence and death of leaves are important in the winter economy of deciduous trees. The 'June drop' of immature fruits is a senescent process which serves to thin out the developing fruit of some trees.

In view of its functional significance, it would appear unlikely that senescence is merely a random disorganisation of the complex activities of mature tissues. Experimental observations support the view that senescence, at least in its early stages, is a positive response, partly to external factors such as daylength, temperature, nutrient supply and also to internal factors such as size, degree of flowering or ripeness of fruits – as already mentioned on page 99, gardeners often remove immature fruits to prolong active growth of other organs (usually flower buds); this shows that ripening fruit may in some way accelerate senescence of other parts of the plant.

Senescence is particularly easy to follow in leaves, as destruction of chlorophyll, shown by progressive yellowing, occurs before other symptoms such as wilting and shrivelling can be seen. Analysis of the biochemical constituents of leaves reaching the end of their mature life shows that the proteins start to breakdown to amino acids, and that this is accompanied by degradation of nucleic acids, presumably into their monomers also, as well as by chlorophyll destruction. Metabolic processes such as respiration and photosynthesis decline, although there may be a short period in the early stages of senescence of an outburst of respiratory activity called the climacteric (figure 13.2); this is particularly noticeable in fleshy fruits like apple and pear.

When a healthy leaf is detached from a plant, senescent

symptoms start immediately in the leaf, showing that in this case the rest of the plant must have been preventing the process. But not all the control comes from the parent plant, as detached leaves senesce more slowly in light than in darkness, showing that part of the control mechanism is in the leaf itself and that it is still responsive to external factors.

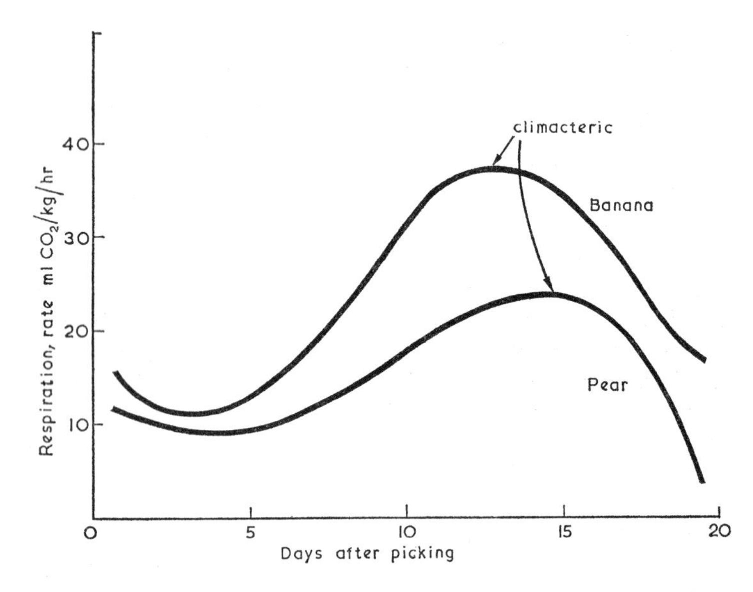

Fig. 13.2. Change in respiration shown by some fruits during ripening (senescence). There is an initial decrease followed by a climacteric rise, followed again by a decrease. (after Leopold)

Leaf abscission

This is often considered under the heading of senescence, but we have already described it in Chapter 7. We can note here that abscission is an active and complex process and not merely a question of death and disorganisation of the petiole – this would result in a dead leaf but not a detached one, as dead petiolar tissue is strong enough to hold a leaf on.

The role of hormones

Although the control of senescence is not yet understood it is likely that hormones are involved, as their application to plant tissues can markedly affect the process.

Auxin may accelerate abscission, especially if applied to the tissue on the proximal (plant) side of the abscission layer. Recently a compound called abscisic acid (figure 4.9) which can also promote abscission, has been isolated from plant tissues.

Kinetin has a dramatic effect; leaves treated with this lose their chlorophyll and their protein at a much slower rate than untreated leaves, remaining green for long periods – indeed, this property has been used commercially to prolong the storage life of green vegetables. Kinetin appears to act by causing mobilisation of amino acids, carbohydrates and other cellular constituents towards the place where it is applied (Plate 16), these being withdrawn from adjacent untreated tissues. This mobilisation effect may develop in the plant because of changed hormone levels in response to an external factor such as daylength or lowered temperature. We have seen in Chapter 4 that roots synthesise cytokinins and it has been known for some time that rooted leaves senesce more slowly than detached leaves without roots, showing that a preventive factor moves from the root to the leaf. We can infer, on this evidence, that it is a cytokinin.

Application of gibberellin or auxin to leaves also inhibits their senescence in some species, and this may be due to the ability of these chemicals to stimulate the synthesis of certain proteins.

Summary

It seems likely that senescence, as with most growth phenomena, is caused and controlled by the complex interaction of many factors, both external and internal. Withdrawal of important metabolites by distant growth regions may be an important factor in organ senescence, and the process can often be delayed by cutting off these regions, for instance, the meristems or the fruits. This would not explain senescence of the whole plant, however. Hormones are also likely to play a direct role in cellular changes, and as senescence starts as an active process they may be expected to do this via the protein synthetic system.

Senescence is now an active field of research. It is a major manifestation of life and, in relation to man himself, is not merely of biological but also of sociological importance. A more definitive understanding of it can be expected in the next few years.

14

General conclusions

We have now studied the various quantitative and qualitative changes that occur in the life of the plant, i.e. its growth and development.

Throughout the preceding chapters we have been concerned with mechanisms of growth processes rather than with descriptions of them. We have sought to understand, for example, how cell expansion or dormancy or flowering are controlled. Now, the growth and development of the plant is, of course, a mirror to all its physiological activities. The final height, form and level of development are ultimately determined by the sum total of events which occur within the organism, namely every aspect of metabolism and of exchange of material between cell and cell and organ and organ. Although we have not dealt specifically with any of these (which are considered in detail elsewhere) we must be careful not to exclude them from our considerations. Even so, one of the major concepts to emerge is that control over growth is exerted by certain chemicals, the phytohormones; we have seen that the sequence of changes undergone by the cell – division, expansion, differentiation – are at some points regulated by cytokinins, gibberellins and auxins, acting perhaps in sequence or sometimes simultaneously. Moreover, we have attempted to demonstrate that the integration of the plant's activities is in many cases achieved by the interaction of these hormones in different parts of the organism.

In the last analysis, of course, growth and development depend upon the environment which acts upon the internal processes. We can appreciate immediately that this is true in a broad sense, for we can easily recognise that the plant obtains its very substance (energy, carbon dioxide, oxygen, mineral nutrients and water) only from its environment. But we have also seen that environmental factors such as light and temperature can have a highly specific effect on growth. Thus, stem elongation, cell differentiation and morphogenesis are all regulated by light of particular wavelengths and sometimes of particular durations. In many instances we know that the environment acts through the intermediation of hormones. This is demonstrated rather nicely in the case of dormancy: short days seem to promote the synthesis of inhibitors in the leaves which then induce bud dormancy, while in seeds cold temperatures might well lead to increased gibberellin production thus breaking the dormant period. Similarly, there is good evidence for the synthesis of specific flowering hormones during photoperiodic 'induction'.

The hormone concept seems to dominate current views on growth control, but we must be careful not to allow it to exclude other ideas. There is much which is not explicable in hormonal terms, such as the action of the phytochrome system, most of morphogenesis and, in many plants, even cell growth itself. As we have emphasised above, the character of the organism is the result of all its physiological activities and to ascribe the whole of growth control to a few specific chemicals is certainly too simple a view. We know, for example, that many unspecific metabolites can profoundly influence differentiation and morphogenesis; sucrose concentration can determine the differentiation of callus into xylem elements, or can cause tuberisation in potato stem segments, to quote but two examples.

Perhaps the major problem to be solved is how the control systems actually work; for example, how do the hormones control the metabolic events of the cell? We are just beginning to appreciate the subcellular and molecular levels at which they might act, and we see that one function may be to regulate protein synthesis. The point at which this regulation might occur is highly debatable, and we have no really solid or conclusive evidence to show which parts of the nucleic acid to protein, transcription–translation mechanism are affected. This is clearly a highly complex problem which, as we have previously stated, lies at the heart of developmental biology. But we are now at exciting times and the plant physiologist together with other biologists is beginning to explore these complicated fields of research.

Molecular biology offers us tantalising prospects but we should not allow these to obscure the fact that we are seeking to understand how the *whole* organism works, not just the cell or the organ. We can only fully understand the growth of the whole plant when we have learned more about the influences that the parts exert upon each other, how they interact, and how they are interdependent.

Those, then, are the tasks which the developmental physiologist has set himself. We hope that in this book we have succeeded in introducing the reader to the present state of our knowledge and to the problems which still lie ahead.

Suggestions for further reading

General

LEOPOLD, A. C. *Plant Growth and Development.* McGraw-Hill, New York, 1964

Certain aspects of growth physiology not dealt with in the present text can be found in detail in this book.

STEWARD, F. C. *Growth and Organisation in Plants.* Addison-Wesley, Reading, Mass. 1968

A scholarly and individual account by an eminent plant physiologist; especially useful for those wishing to delve deeply into the subject.

TORREY, J. G. *Development in Flowering Plants.* Collier-Macmillan, London 1967

A good general account of development, especially from the viewpoint of morphogenesis.

Morphogenesis and differentiation

CUTTER, E. G. (Ed.) *Trends in Plant Morphogenesis.* Longmans, London 1966

An advanced survey with contributions from several authors who are working actively in the subject.

WADDINGTON, C. H. *Principles of Development and Differentiation.* Collier-Macmillan, London 1966

A very readable and authoritative book dealing with the fundamental problems of differentiation, looked at by an embryologist and geneticist.

Plant hormones

AUDUS, L. J. *Plant Growth Substances.*

A standard work, mainly on auxins, for the specialist. Contains much information about the practical aspects of growth substances.

VAN OVERBEEK, J. 'Plant hormones and regulators', *Science*, **152,** pp. 721–31, 1966

A stimulating short review paper summarising current views on hormonal control, including the interactions of hormones in the plant: by a well-known growth physiologist.

Flowering

HILLMAN, W. *The Physiology of Flowering*. Holt, Rinehart and Winston, New York 1964

A good account of the subject by an active worker in the field of photomorphogenesis.

SALISBURY F. B. *The Flowering Process*. Pergamon Press, Oxford

A rather detailed work written for the potential researcher, but well worth the attention of the beginner.

In addition, authoritative review papers covering aspects of growth and development appear in *Annual Reviews of Plant Physiology*, Annual Reviews Inc. Palo Alto.

Index